SPIRULINA
The edible micro-algae of ancient Origins

Ifeanyi Charles Okoli

COPYRIGHT © 2024 OKOLI, IFEANYI CHARLES

All rights reserved.

No part of this publication covered by the copyright herein should be reproduced, stored in a retrieval system, or be transmitted in any form, or by any means graphic, electronic, or mechanical, including but not limited to photocopying, recording, scanning, digitizing, taping, Web distribution, information networks, or information storage, without the express written permission of the author or publisher, with the exception of brief excerpts in magazines, articles, reviews etc. Please purchase only authorized electronic edition, and do not participate in or encourage electronic piracy of copyrightable materials.

Published by Tapas Publishing Co
Owerri, Nigeria

Cover Picture: A Kanembu woman processing spirulina near a lake in the Republic of Chad, Central Africa. Image source: Robert Henrikson

DEDICATION
To the Kanembu women of Chad

CONTENTS

Preface

The Spirulina Algae
    Introduction
    The different uses of spirulina
    The market for spirulina products
    Conclusion

An Ancient Blue-Green Micro-Algae and the Best Food of the Future
    The history of spirulina
    New initiatives for spirulina cultivation
    Commercial production of spirulina
    Conclusion

The Commercial Production of Spirulina
    Biological characteristics of spirulina
    Small-scale production of spirulina
    Commercial spirulina production
    Conclusion

Biochemical Compositions of Spirulina
Introduction
Physical properties of spirulina
The protein content of spirulina
The lipid content of spirulina
The carbohydrate content of spirulina
Vitamin and mineral content of spirulina
Conclusion
The Food Value of Spirulina
Introduction
    Improvement of the nutritional value of common foods

Spirulina in vegetarian and vegan diets
Spirulina in sports nutrition drinks
Consumer acceptance of spirulina
Nutritional safety of spirulina products
Quality control of spirulina products
Conclusion

The Health Benefits of Spirulina Algae-1
Spirulina as a nutritional supplement
The effect of spirulina on intestinal flora balance
Detoxification effects of spirulina
Anti-microbial effect of spirulina
The anti-viral effects of spirulina
The effect of spirulina on heavy metal toxicity
Conclusion

The Health Benefits of Spirulina Algae-2
Introduction
Regulation of immunity
The role of spirulina in the management of obesity

The role of spirulina in reduction of blood cholesterol level
The role of spirulina in lowering the risk of diabetes
Anti-inflammatory activities of spirulina
Anti cancer properties of spirulina
The role of spirulina in the treatment of cardiovascular disease
Anti-cancer properties of spirulina
Effect of spirulina on aging
The role of spirulina in the treatment of neurodegenerative disorders

Uses of Spirulina in Animal Production

SPIRULINA AS ANIMAL FEED SUPPLEMENT

Effects of dietary spirulina supplementation on poultry performance
    Effects of dietary spirulina supplementation on pig performance
    Effects of dietary spirulina supplementation on ruminant performance
    Effects of dietary spirulina supplementation on rabbit performance
    Effects of spirulina supplementation of aquaculture feeds
    Use of spirulina in pet care
    Conclusion
Uses of Spirulina in Crop Production
Spirulina as a natural bio-fertilizer
Spirulina as a plant bio-stimulant
Spirulina as a bio-pesticide
Conclusion
Environmental and Other Uses of Spirulina
Introduction
The use of spirulina in bioremediation of polluted sites
Production of biofuel from spirulina
Production of bioplastic from spirulina
Production of dyes from spirulina
Conclusion

PREFACE

Spirulina is an edible algae with a diverse range of food and health applications, especially for improved digestion, energy, weight management, mental clarity, and immune functions. Spirulina has therefore been promoted as having major advantages in the fight against protein-energy-malnutrition, protein-energy-wasting, and malnutrition resulting from micro-nutrient deficiencies. Spirulina has been associated with health benefits such as anticancer, anti-diabetic, anti-inflammatory, immunomodulatory, anti-aging and anti-anemic effects. The algae has also been successfully applied in the bioremediation of wastewater, as well as in cosmetics, livestock, poultry, crop farming, aquaculture, bioplastics, and biofuel production. It has also been proven that spirulina can help to reduce greenhouse gas emissions through its ability to fix CO_2. Several ongoing innovative research on the applications of spirulina points to its potential roles across multiple sectors, thus making it an emerging natural eco-friendly resource. This book presents a body of information on the current uses of the spirulina microalgae in many industries. The book is written in a simple, and understandable format, with the goal being to simplify the available technical research information on the subject for the benefit of the general public. It is hoped that the readers will find the information contained in this book useful, both for business and daily activities.

THE SPIRULINA ALGAE

*(Algae growing on rocks at the bank of a lake
Image source: www.csfhungary.hu)*

Prof Ifeanyi Charles Okoli

Introduction

Spirulina is one of the oldest lifeforms on earth. It is a blue-green algae that traditionally grows in both fresh and saltwater bodies, and has been confirmed as a rich source of proteins and micro-nutrients. Spirulina grows naturally in high salt alkaline water reservoirs in subtropical and tropical Africa, America, and Asia. It thrives well in warm, sunny environments, reproducing rapidly, and forming thick greenish mats on the water surface. Hot springs having temperatures as high as 35 to $45^\circ C$, also provide a highly alkaline environment ideal for spirulina growth. Some natural springs containing high concentrations of minerals, also support the growth of several blue-green algae including spirulina. Again, spirulina grows naturally in rice paddies, where the nutrient-rich water provides an ideal environment for the algae to grow and reproduce. In all these environments, spirulina can be harvested and processed for several applications, although it can also absorb some toxins or pollutants, making it unsafe for consumption or other uses. In recent times artificial environments have been developed for experimental and commercial production of spirulina in many regions of the world.

Spirulina is an edible algae with a diverse range of food and health applications, especially for improved digestion, energy, weight management, mental clarity, and immune functions. It is regarded as a superfood because of its content of between 60 and 70 percent protein, eight essential, and all the non-essential amino acids, as well as high levels of beta carotene, minerals, and vitamins. Spirulina is also rich in phycocyanin, a pigment-protein antioxidant complex found mostly in blue-green microalgae. Spirulina has therefore been promoted as having major advantages in the fight against protein-energy-malnutrition, protein-energy-wasting, and malnutrition resulting from micro-nutrient deficiencies. Its health potential and the ease with which it can be grown locally have therefore been highlighted in recent times. For example, while very little space, water, and time are needed to achieve its high yield per unit area, the amount of protein in the algae requires that only about 1 to 3 grams per day for 4 to 6 weeks is sufficient to rehabilitate a malnourished child. It has also been proven that spirulina can help to reduce greenhouse gas emissions through its ability to fix CO_2. Consequently, several start-up and small-medium enterprises have been initiated or promoted by multi-national organizations and NGOs in the developing countries to promote micro-algae and spirulina cultivation. Commercially, spirulina is produced in large outdoor ponds under controlled conditions or directly from lakes and processed as dried biomass.

The Different Uses of Spirulina

In addition to being considered a superfood because of its high nutrient content, spirulina has several other beneficial uses. It has specifically been approved as human food by governments, health agencies, and associations in many countries, and is therefore marketed and consumed in these countries. It is increasingly being used as a dietary supplement because of its rich protein level of up to 70 g per 100 g dry weight, superior to meat and legumes. Recent studies have also unlocked the potential additive benefits of spirulina in pet foods and livestock feed. It is a cost-effective additive for sustainably improving livestock and crop productivity to ensure food and nutritional security in several developing countries. Sow and Ranjan reported that spirulina has been associated with health benefits such as anticancer, anti-diabetic, anti-inflammatory, immunomodulatory, and anti-anemic effects. It has also been successfully applied in the bioremediation of wastewater, as well as in cosmetics, organic farming, aquaculture, and biofuel production. Novel methods are currently being employed to improve the extraction and application of different products from spirulina as shown in Figure 1.

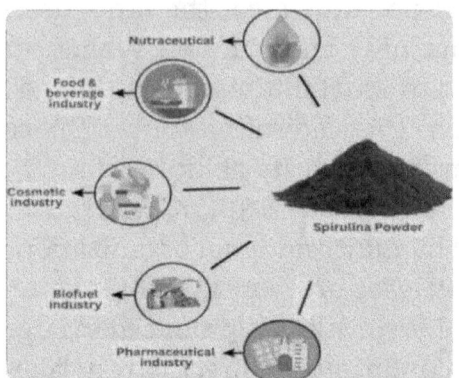

Fig. 1: Applications of spirulina extracts (Source: Kalid et al., 2024)

The Market for Spirulina Products

According to the Acumen Research and Consulting GlobeNewswire 2022, the global spirulina market size was USD 640 million in 2024 and has been projected to reach USD 1.20 billion by 2031 at a cumulative annual growth rate (CAGR) of 9.4 percent. Dataintelo (2022) however reported a higher USD 2.5 billion global market size in 2021, which was projected to reach USD 5.4 billion by 2030 at a CAGR of 10.1 percent. In terms volume, the market is projected to reach 102,381.3 tons by 20230 at a CAGR of 8 percent. North America currently accounts for 35 percent of the market. Specifically, the United States of America algae protein market in 2021 alone was USD 2.1 million, projected to grow at 10.6 percent up to 2030 (Figure 2), indicating that the USA is the most important market for the product.

This market growth has been driven by the high demand for fresh/frozen spirulina by the more than 80 million vegetarians globally, who have adopted it as a major source of protein and other nutrients. It is also a protein supplement for the bodybuilding and sports industry. Spirulina is currently presented in the market as powder, tablet, and extracts, with the powder being the most widely used, either in pre-packed meals and food products or mixed with drinks or sprinkled on foods to improve their nutritional values. The powder is also presented as tablets and energy bars, and sold in pharmaceutical or health food stores as oral supplements. The inclusion of spirulina extracts in natural solutions, and other health products has also gained attention as body purifiers and energy/immune system boosters.

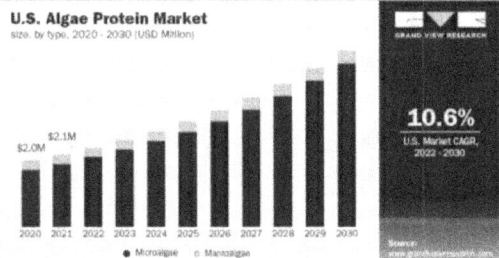

Fig. 2: *The United States of America algae protein market and projections*

Spirulina as feed for livestock, poultry, and aquaculture has been shown to improve their yields, and product quality. Its use as feed additive has therefore increased in recent times. In this regard, North America accounts for about 25 percent of the market, followed by Europe and Latin America. Currently, Japan, the United States of America, and the European countries are the main importers of Mexican spirulina powder. Vitamins A and C are added to Lozenges and capsules made from the powder following the regulations of the importing countries. The crude powder is also exported to the United States of America, while the lozenges and capsules are sold almost exclusively in dietetic shops.

Conclusion

Spirulina microalgae grows in many tropical developing countries, especially in sub-Saharan Africa. Its production could be used to mitigate the prevalent malnutrition and micronutrient deficiencies among vulnerable groups in these countries. Although the current major global use of spirulina remains in food and nutrition, new studies point to its value in health promotion, agriculture, biofuel, cosmetics, and other industrial and environmental applications. These novel applications could potentially enhance the already expanding global market for green algae.

AN ANCIENT BLUE-GREEN MICRO-ALGAE AND THE BEST FOOD OF THE FUTURE

A Kanembu woman harvesting spirulina in Lake Chad (Image source: www.fishconsult.com)

Prof Ifeanyi Charles Okoli

The History of Spirulina

Spirulina is a blue microalgae believed to be one of the earliest life forms on earth, dating about 3.5 million years. Human utilization of spirulina for food and other benefits has been traced back to the ancient Aztecs and ancient Mayan cultures, who harvested the algae from lakes and other water bodies. Sánchez and colleagues reported that in the 16th century, the Spanish conquistador Hernán Cortés documented the harvesting and consumption of the algae by the Aztecs. Although it has lost its popularity as a food source among the peoples of these cultures, it is still culturally consumed in several parts of Africa and Asia. Spirulina has been traditionally consumed by various indigenous populations in Africa, particularly in Chad, where it grows naturally in Lake Chad and other water bodies of the arid region. The largest spirulina lakes are indeed found in Central Africa around Lakes Chad and Niger, and in East Africa along the Great Rift Valley where it may be one of many algal species. Lakes Bodou and Rombou in Chad have been reported to harbor a stable monoculture of spirulina dating back centuries. It is also a major species in Kenya's Lakes Nakuru and Elementeita, and Ethiopia's Lakes Aranguadi and Kilotes.

Aztecs harvesting blue-green algae from lakes in the Valley

of Mexico. Drawing in Human Nature, by Peter T. Furst, 1978)

According to the reports by Habib and colleagues, in 1940 the French scientist, Dangeard observed the consumption of spirulina cakes called dihé by the Kanembu people near Lake Chad in the present-day Republic of Chad in Central Africa. He also identified the algae in lakes in the Rift Valley of East Africa, where it served as the main food for the flamingos living around those lakes. Similarly, the Belgian scientist, Jean Léonard reported that the edible spirulina cakes obtained from villages near Lake Chad were being sold in native markets at N'Djamena the capital of Chad. In 1967, the International Association of Applied Microbiology established spirulina as a "wonderful future food source" and today it is widely cultured throughout the world.

New Initiatives for Spirulina Cultivation

In recent years, there has also been an increased interest in the cultivation and consumption of spirulina in Africa due to its high nutritional value, and potential for addressing malnutrition and food insecurity. Overall, the cultivation and consumption of spirulina in many developing countries have the potential to contribute to improved nutrition and food security, especially in regions that are prone to environmental challenges. Several international and national organizations promote the cultivation and consumption of spirulina, particularly as a source of protein, and other essential nutrients in Africa. For example, the "Spirulina Initiative" has been working in Chad since 2005 to promote the cultivation of spirulina by local communities, through the provision of training and support for small-scale spirulina farmers, and the promotion of the use of spirulina as a nutritional supplement for children, and pregnant women. Other initiatives that involve partnerships between local organizations, universities, and international agencies have also promoted the consumption of spirulina as a source of quality nutrients, and income for local communities in Kenya, Tanzania, and Burkina Faso.

Modern processing of spirulina in Chad. (Image source: Kopeprod)

Spirulina production and consumption however remains relatively unpopular in most sub-Saharan African countries due to several constraints such as lack of effective promotion, high production cost, low production capacity, odor and taste issues, and limited consumer awareness about its many benefits. Again, commercial spirulina production is usually carried out in a controlled environment using photo-bioreactors that require significant initial capital input, which may be too expensive for small-scale local farmers. Piccolo and Short however reported that spirulina continues to grow naturally in many lakes in sub-Saharan Africa, especially in Kenya, Ethiopia, Sudan, Tanzania, Botswana, and Chad, and is being harvested by the locals. Spirulina is also found in soils, marshes, brackish water, seawater, and thermal springs. In these natural water bodies, the algae population grows rapidly, reaches a maximum density, and then dies off when the nutrients are exhausted. The new growth cycle begins when the decomposed algae release nutrients into the water body. Nutrient supply is therefore renewed through either tidal wave upwelling of the water bodies, or influxes of nutrients from streams and rivers that flow into them. The growth cycle of the algae is therefore regulated by the limited supply of nutrients in these natural water bodies.

In Chad, artisanal production of spirulina cakes called dihe persists according to the ancient methods. The women harvest the spirulina from the lake by skimming the water surface with locally produced nets. They then place the sludge into a hole dug in the warm sand to drain the water while leaving behind a round cake-like dry substance. The cake is cut into slices and sold at local markets to consumers who use it to improve the protein content of their mostly grain-based diets. A slightly modified more modern artisanal method of harvesting and processing spirulina from its natural environment was however recently described by Tidjani and colleagues. In this method, the women enter the water body and collect the water containing the spirulina algae with plastic cups or buckets. The water is then passed through a 5 mm sieve mesh which retains the spirulina and some impurities. The filtrate is further passed through a second 600 μm mesh screen which retains spirulina and allows a large part of the water to pass through. The spirulina is then collected with a cloth and expressed to remove most of the water. The remaining spirulina pest or cake is dried in a solar dryer for 4 to 5 hours. The dried product may be cut into smaller pieces or ground into powder and packaged for sale. Such a product has been reported to be of better hygienic quality, indicating the need to promote the approach among local spirulina producers and processors.

Commercial Production of Spirulina

Spoehr and Milner (1949) were the earliest promoters of the mass culture of algae that would help to overcome global protein shortages, although their work received only casual interest. The first systematic and detailed study of the growth requirements, and physiology of spirulina was however carried out by Zarrouk (1966) and established the basis for the first large-scale production plant of spirulina. The first commercial spirulina processing plant was however set up in 1969 by the French company, Sosa Texcoco Ltd. The commercial business was however established in 1973. The company harvests spirulina from an area of Lake Texcoco, at 2 200 m above sea level, in a semitropical environment, in the Valley of Mexico, with an average annual temperature of 18 ºC, and is the largest single plant for the production of spirulina biomass. The commercial plant employs a semi-natural cultivation process which consists of harvesting during day and night, and inducing the algal biomass to double in three to four days. The initial production capacity of 150 tonnes of dry spirulina biomass per year was later increased to 300 tonnes per year, produced from 12 hectares of natural ponds. After filtration, homogenization, and pasteurization, the algal biomass is spray-dried. In Myanmar, Asia, the production of spirulina in the semi-natural Twin Taung Lake was also initiated in 1984 and by 1999 was producing 100 tonnes/year, with about 60 percent of the biomass being harvested from the surface of the lake, and about 40 percent grown in outdoor ponds alongside the lake. The spirulina is harvested on parallel inclined filters, washed with fresh water, dewatered, pressed again, and the paste extruded into noodle-like filaments, and dried in the sun on transparent plastic sheets. The dried products are then supplied to a pharmaceutical factory that pasteurizes and presses them into tablets.

(Drying of process spirulina noddles in Chad. Image source: Kopeprod)

Although Sosa-Texcoco stopped production of spirulina in 1995, the quest to understand the organism better has continued, and has given rise to numerous research activities, and the establishment of numerous commercial production units even at places where the organism does not grow naturally, thus reflecting its perceived global importance. Over the years, the promotion of spirulina as a human health food was associated with controversial claims of numerous health benefits such as the cure of specific cancers, and antibiotic and antiviral activities which were at that time not backed up by detailed scientific, and medical research. Interest in spirulina as a potential solution to world hunger has continued to grow and has led to its large-scale cultivation in many high-income countries. Again, spirulina gained some popularity after it was successfully used by the National Aeronautics and Space Administration (NASA) as a dietary supplement for astronauts exposed to high levels of oxidative stress, redox imbalance, and elevated inflammation due to extra- and intra-cellular activities resulting from reduced gravity, ionizing radiation, and variability in atmospheric conditions during space missions. Spirulina is now consumed all over the world as a health supplement, and ingredient in a variety of foods, from smoothies to energy bars for vegetarians, vegans, athletes, and health enthusiasts. Villaro-Cos and colleagues have also enriched several food products such as beverages, biscuits, dairy, and bread with spirulina. It was declared the world's first superfood, with a complete nutrient profile, and best food of the future" for its eco-sustainability, and also listed by the US Food and Drug Administration (FDA) in the category of generally recognized as safe (GRAS).

Conclusion

Human utilization of spirulina for food and other benefits is an ancient practice of several cultures in Africa, Asia, and the Americas. The increasing artisanal and commercial production of spirulina in recent times, even at places where it does not grow naturally, reflects its perceived global importance. Its complete nutrient profile, and current global acceptance as a health supplement by vegetarians, vegans, athletes, and health enthusiasts support its promotion as the best food for the future.

Therefore, there is a need for local entrepreneurs to exploit the economic benefits of this eco-sustainable ancient blue-green microalgae.

THE BIOLOGICAL CHARACTERISTICS AND COMMERCIAL PRODUCTION OF SPIRULINA

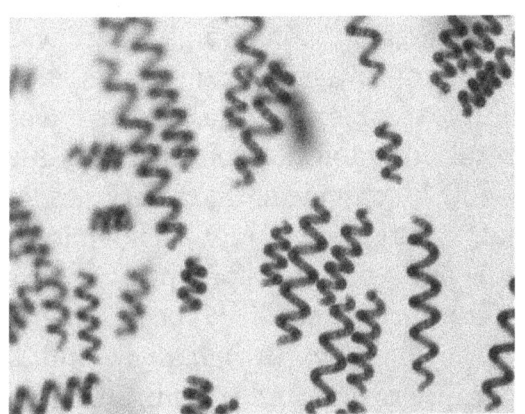

Microscopic view of Spirulina (Image source: Koru, 2012)

Biological Characteristics of Spirulina

Spirulina (*Arthrospora*) is a photosynthetic, filamentous, spiral-shaped, multi-cellular, and blue-green micro-alga that was initially classified under the plant kingdom due to its rich plant pigments, and its ability to photosynthesize, but was later included in the bacterial kingdom (*Cyanobacteria*) due to its genetic, physiological, and biochemical characteristics. It is a symbiotic bacteria that could be rod- or disk-shaped, and fix nitrogen from air. Several varieties of spirulina are recognized, with the most commonly studied species being *Spirulina platensis* (*A. platensis*), *S. maxima* (*A. maxima*), and *S. fusiformis* (*A. fusiformis*). According to Sow and Ranjan, spirulina is morphologically a multi-cellular, filamentous, unbranched trichome made up of cylindrical and spiral loose cells containing gas-filled vacuoles which along with the helical shape of the filaments makes it float as a mat on the water surface. The helical shape of the filaments is maintained only in a liquid environment or culture medium. Although the helical shape of the trichome is characteristic of the genus, the helical parameters vary in different species, and even within the same species. The cylindrical filament shape measures about 50 to 500 μm in length and 3 to 4 μm in width. Spirulina has a cell wall that contains peptidoglucan, a lysozyme-sensitive heteropolymer similar to the cell wall of Gram-negative bacteria. The cell wall confers shape, and osmotic protection to the cell, in addition to other materials not sensitive to lysozyme.

Spirulina reproduces asexually by cell division, to produce two identical daughter cells, although sexual reproduction can occur under certain conditions. Ali and Saleh reported that the four fundamental stages, in the life cycle of the spirulina are trichome fragmentation, hormogonia cell enlargement, maturation processes, and trichome elongation. Spirulina grows naturally in freshwater environments such as lakes, rivers, ponds, and salty waters. It requires bright light for photosynthesis, with the optimal light intensity ranging from 5000 to 7000 lux. Its photosynthetic activities produce chlorophyll, which gives it its green color, and phycocyanin, which gives it its blue color. The main photosynthetic pigment is however phycocyanin. The optimal temperature for its growth and cultivation ranges from 25 to 35°C, and slightly alkaline pH between 8.0 and 11.0, with optimal growth occurring at 9.0 to 10.0. Spirulina requires a variety of nutrients such as nitrogen, phosphorus, potassium, iron, zinc, and manganese for growth, with the ideal nutrient ratio being about 10: 1: 0.1 (N: P: Fe respectively). Alkaline, saline water (>30 g/l), with high pH (8.5 – 11.0) favors good production of spirulina, especially where there is a high level of solar radiation at high altitude in the tropics. The ability of spirulina to utilize ammonia as a source of nitrogen at these high alkaline pH levels has been attributed to its relatively high cytoplasmic pH of 4.2 to 8.5.

Different strains of spirulina may however have specific nutrient requirements, and the conditions for their cultivation and growth. For example, the higher the pH and the conductivity of the water, the greater the likely predominance of spirulina species over other alga species. *S. platensis* and *S. maxima* thrive in highly alkaline lakes of Africa and Mexico where the cyanobacteria population is practically monospecific. As an obligate photoautotroph, spirulina cannot grow in the dark on media containing organic carbon compounds. It can only utilize carbon dioxide in the presence of light, and utilizes mostly nitrates as a nitrogen source to produce glycogen through photosynthesis. Other specific characteristics of spirulina include the ability of the thermophilic or thermo-tolerant strains to thrive at temperatures between 35 and 40 °C, a property that confers the advantage of eliminating microbial mesophilic contaminants. Spirulina has also been shown to be highly resistant to ultraviolet rays. Soni and coworkers however reported that in natural habitats, the spirulina growth cycle is regulated mostly by the supply of nutrients, with higher growth rates being observed at higher CO_2 fixation efficiency, and supplementation of the water medium with high-value products rich in organic matter. Other studies have also shown that spirulina growth and nutritional composition are influenced by variable environmental conditions such as light intensity, temperature, and salinity, as well as nutrient availability and growth phases. Generally, the photosynthetic process performed by micro-algae as a primary producer of oxygen contributes to fixing the carbon and nitrogen which in turn increases the algae cells produced with high value of proteins and lipids. The three major light-harvesting pigments in spirulina are chlorophylls, phycobilliproteins (especially phycocyanin which is the most abundant), and carotenoids. These pigments aid the synthesis of several important enzymes that benefit human, and animal metabolic processes.

Small-Scale Production of Spirulina

Spirulina indeed lends itself to simple technology suited for small-scale production. It has been promoted as a potential income-generating activity for households or community cooperative societies and local consumers, especially in communities where poor dietary regimes need supplementation. Extensive or semi-intensive spirulina production for use as animal or aquatic feed has also been used to reduce the cost of production for small-scale farms and aquaculture. FAO (2008) reported that spirulina production could be carried out in unlined ditches of low flow of 10 cm/second, while stirring may be provided by a simple wind-driven device or by manual turning. A suitable cloth could be used for harvesting the biomass which is then dewatered by manual pressing before drying in the sun. Although the quality of the spirulina product obtained from such small-scale operations has been reported to be lower than those obtained from the hi-tech clean cultures, it could serve as quality animal feed. Researchers have also designed and demonstrated the integration of village-level waste management, biogas generation, spirulina production, composting, and fish culture for developing countries. In a Bangladeshi study with limited success, a digester was used to process sewage and other wastes to produce biogas for community cooking facilities, while the liquid effluent was treated in a solar heater and then used to fertilize algal ponds. The algal biomass was harvested with woven cloths and dewatered with solar driers.

A much better design by the Murugappa Chettiar Research Centre in Chennai, India has however been successfully introduced on a large-scale in the rural communities of the district of Tamil Nadu. The design involved the use of a mud pot spirulina production approach consisting of a medium made of biogas slurry, 2 – 3 g of sea salt, or a combination of potassium dihydrogen phosphate, cooking soda, and sodium chloride in place of the sea salt, and pure spirulina culture. The mud pot was buried up to its neck in the ground, filled with water mixed with the medium, and thereafter inoculated with a small quantity of the pure spirulina culture. The inoculated medium was exposed to sunlight, with intermittent stirring of 3 to 4 times a day to ensure optimal growth of the spirulina to maturity within 3 to 4 days. The mature spirulina was harvested by a simple cloth-filtration method, washed with fresh water to remove impurities, and used in preparing noodles, pulses, or vegetables. It could also be preserved by sun-drying immediately after harvest to maintain quality.

Several studies have shown that the growth medium is the most important cost item in small-scale spirulina production, and therefore should be kept as down as possible to make the product cost-effective. One of the effective methods of achieving this is by using locally available and cheap organic waste effluents such as pig slurry, or effluent from a fertilizer company mixed with seawater at 50 percent dilution to achieve a pH of 8.5 after 21 days. Other organic waste effluents that have yielded positive results include wastewater from Thai rice noodle factories, rice husk ash, lagoon water, and fish pond wastewater. Specifically, Khoirunisa and coworkers used palm oil mill effluent (POME) as a medium to cultivate *S. platensis* using a raceway open pond bioreactor for 5 days. The study design was, POME concentration (3 × 4 × 5 × diluted), and feed loading concentration of *S. platensis* (0.443 g/L; 0.61, 0.952 g/L) as variables, affecting the spirulina performance in terms of growth, and biomass yield on dry weight basis. The result showed that a five-fold dilution of the POME with water and 0.443 g/L concentration of the organism produced the highest dry weight of the spirulina.

Commercial Spirulina Production

The commercial production of spirulina is based on a proper understanding of the cultural characteristics of the organism. According to Ayala (1998), the major environmental factors that influence the productivity of spirulina are luminosity (photo-period 12/12, 4 luxes), temperature (30 °C), inoculation size, stirring speed, dissolved solids (10 – 60 g/litre), pH (8.5 – 10.5), water quality, and macro- and micro-nutrients (C, N, P, K, S, Mg, Na, Cl, Ca, and Fe, Zn, Cu, Ni, Co, Se).

(Image source: Earthrise Farms by Chris Yoro and Rodney Ruppert)

Spirulina could be cultured in different types of media, especially inorganic and decomposed organic nutrients. Bhattacharya and Shivaprakash however reported that *S. platensis* records higher growth rate, biomass yield, pigment concentration, and low intracellular phenolics under cultural conditions than other spirulina species. Sanchez-Luna and colleagues in their studies reported that intermittent inclusion of urea as a nutrient source could be used to reduce the production costs of spirulina in large-scale facilities. Bangladeshi studies have also demonstrated cost-effective spirulina cultures agro-industrial wastes such as sugar mill effluent, poultry wastes, fertilizer factory waste, organic urban waste, and several other organic biomass wastes.

The two major approaches to the commercial cultivation of high-value spirulina are presently the closed photobioreactors (PBR), and the open pond method, with *S. platensis* and *S. maxima* being the two species mostly cultivated due to their valuable components, positive and nontoxic effects on humans. The production of spirulina requires careful monitoring and control of various environmental factors to ensure optimal growth and quality. The first step should be the preparation of the culture medium, which should contain the appropriate nutrients in the right proportions. The culture medium should typically consist of water, nitrogen, phosphorus, and a carbon source such as glucose or molasses. The second step is the medium inoculation with the required amount of the desired spirulina species to initiate growth. Cultivation is the third step during which the inoculated medium is transferred to a large pond or tank where the spirulina is exposed to the appropriate temperature and pH levels to ensure optimal growth. The fourth step is the harvesting of the spirulina after it has grown into a dense mat on the surface of the pond or tank. This mat is harvested using a fine mesh net and then washed with water, pressed, and dried to remove excess water. The final step is the packaging of the dried spirulina for sale, either as a food supplement or for other industrial applications. Soni and coworkers illustrated in Figure 3 the flowchart for spirulina cultivation phases starting from strain selection to pellets formation.

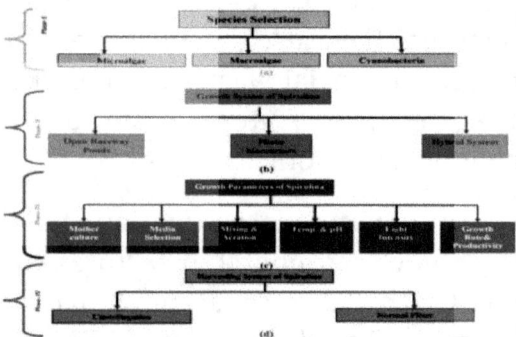

Fig. 3: Different phases of Spirulina cultivation system (Source: Soni et al., 2017)

The composition of the culture medium is a critical component of commercial spirulina production which should be managed carefully since large quantities of inorganic salts such as carbonates and bicarbonates are utilized to achieve an alkaline pH of 9.5 – 9.8 to effectively prevent contamination by other microalgae. The investment in the nutrient medium is therefore usually high, representing the second most important factor influencing commercial spirulina biomass production, and accounts for about 15 to 25 percent of the production costs. Again, commercial large-scale spirulina cultivation is usually carried out in shallow open raceway ponds, equipped with paddle wheels to mix the culture. Two types of open raceway ponds could be lined by concrete thereby making it expensive to construct or just a shallow earthen tunnel lined with polyvinyl-chloride (PVC) or some other durable plastic material. Since the raceway pond construction is a major cost item in commercial algae production, several studies have researched low-cost alternative approaches such as clay sealing. Ideally, commercial raceways should be as big as 0.1 to 0.5 hectares, and have a culture depth of 15 to 18 cm, while the ideal paddle wheel diameter should be about 2.0 m and rotating at a speed of 10 rpm. Smaller paddle wheels of 0.7 m in diameter, rotating at a speed 2 to 3 times faster than the 2.0 m diameter paddle wheel, have also been found effective as stirring devices. If the paddle wheel does not sufficiently create the required turbulence for the optimum light pattern in the pond, other means may be used to increase the turbulence in shallow ponds, and consequently their productivity. In places like Nigeria characterized by a lack of reliable electricity and industrial automation, open bioreactors are recommended for spirulina production.

After cultivation, additional processing steps are employed in the biorefinery to transform the spirulina biomass into a finished product, with the main objective being to reduce the water content to between 10 and 25 percent solids by weight. Ayala (1998) divided these processing activities into initial filtration and cleaning with a nylon filter at the entrance of the water pond.

1. In large-scale commercial operations inclining or vibrating screens may be used to harvest the spirulina filaments. The inclining screen has 380 – 500 μm mesh with a filtration area of 2 – 4 m^2 /unit, and is capable of harvesting 10 – 18 m^3 of spirulina culture per hour, and biomass removal efficiency of up to 95 percent. Vibrating screens could be designed as double or triple decks of screens of up to 183 cm in diameter and have similar filtration efficiency as the inclining screens, although a third of the area/unit. Repeated harvesting is used in some operations to enrich the culture with short filaments of spirulina that readily pass through the screen. The product obtained after filtration is usually 8 – 10 percent of dry slurry which requires further concentrated by filtration using vacuum tables or vacuum belts, depending on the production capacity. Other filtration approaches include chemical filtration methods such as flocculation and coagulation, biological methods such as auto flocculation and flocculation approaches, and charge-based separation.

2. Pre-concentration to obtain algal biomass which is then washed to reduce salt content.

3. Concentration to remove the highest possible amount of interstitial water which is located among the filaments.

4. Neutralization of the biomass by adding acid solution.

5. Disintegration or breakdown of the filaments using a grinder.
6. Dehydration by spray-drying. It is estimated that this processing stage represents between 20 and 30 percent of the production costs of spirulina biomass. The spray dryer and drum dryer methods are commonly used for drying spirulina biomass. The spray dryer achieves higher water evaporation capacity (10,000 kg evaporated water h−1) than the drum dryer (1000 kg evaporated water h−1). Overall, the drum dryer has higher capital and maintenance costs, since more units are required to meet the equivalent evaporation capacity of the spray dryer. The energy cost of operating the drum dryer is however much lower than that of the spray dryer. The choice of drying method, especially at the industrial-scale depends on the type of end-product desired. Other methods of drying spirulina biomass include sun drying, cross-flow drying, vacuum trays, and freeze drying.
7. Packing in sealed plastic bags to prevent the hydroscopic action of dry spirulina
8. Storage in dry, unlit, pest-free, and hygienic storerooms to preserve the spirulina pigment.

According to FAO (2008) report, at least 22 countries, including Benin, Brazil, Burkina Faso, Chad, Chile, China, Costa Rica, Côte d'Ivoire, Cuba, Ecuador, France, India, Madagascar, Mexico, Myanmar, Peru, Israel, Spain, Thailand, Togo, United States of America and Vietnam are involved in some form of commercial spirulina production. Companies such as Earthrise Nutritionals in California, USA, Hainan DIC Microalgae in Haidian Island, China, Cyanotech in Kailua-Kona, Hawaii, USA, Olson Nutrição Ltda in Brazil, and Spirulina Mater in Chile are currently the major industrial large-scale producers and employ the raceway-type bioreactors that cover areas between 1000 and 5000 m². Costa and colleagues reported the different methods employed by these organizations to produce their final products for the market. For example, the Spirulina Mater in Chile uses vacuum filtration to concentrate the biomass, which is then washed to remove the salts from the culture medium, and dried in a spray dryer. Earthrise Nutritionals recovers spirulina biomass by passaging the culture through stainless steel mesh, for rinsing and concentration. The concentrated biomass is then transferred by gravity to vibrating sieves and then to a vacuum belt and spray dryer to produce the spirulina biomass powder. Olson Nutrição Ltda on the other hand, uses filtration through stainless steel screens as its recovery method, followed by extrusion, and oven drying, while powder and desired granule sizes are achieved by ball milling.

Conclusion

Spirulina is a photosynthetic, filamentous, spiral-shaped, multi-cellular, and blue-green micro-alga classified as *Cyanobacteria* due to its genetic, physiological, and biochemical characteristics. The optimal growth and cultivation conditions were abundant sunshine, temperature range of 25 - 35 °C, and slightly alkaline pH range of 8.0 - 11.0. A variety of nutrients such as nitrogen, phosphorus, potassium, iron, zinc, and manganese are also required for its growth in both natural commercial environments.

Spirulina production could serve as a potential income-generating activity for households or community cooperative societies and local consumers, especially in communities where poor dietary regimes need supplementation. It can also be produced for use as a cost-effective bio-fertilizer, protein feedstuff for livestock, poultry and aquaculture. Commercial spirulina cultivation is currently photobioreactors, or by the open pond method, with *S. platensis* and *S. maxima* being the mostly cultivated species. There are increasing numbers of commercial spirulina units in several African, Asian, European and American countries, highlighting the growing global popularity of the algae.

Prof Ifeanyi Charles Okoli

BIOCHEMICAL COMPOSITION OF SPIRULINA

Spirulina algae powder (Mage source: Bodi4Life)

Introduction

As cyanobacteria, spirulina species are widely studied for their nutritional and medicinal properties. Spirulina is rich in nutrients, such as proteins, vitamins, minerals, carbohydrates, and fatty acids. It is generally low in carbohydrate content, with most of its composition being protein and fats. Spirulina contains about 60 –70 percent protein by dry weight as shown in Figure 4, representing the highest protein of any natural food, far more than the protein content of animal and fish flesh, soybeans, dried milk, peanuts, eggs, grains or whole milk. The biochemical composition of spirulina grown either under laboratory conditions, collected in natural conditions, or a mass culture system using different agro-industrial waste effluent varies in response to the salinity of the growing medium. Vonshak and coworkers specifically reported that salt-adapted cells had a modified biochemical composition with a reduced protein, and chlorophyll and increased carbohydrate content.

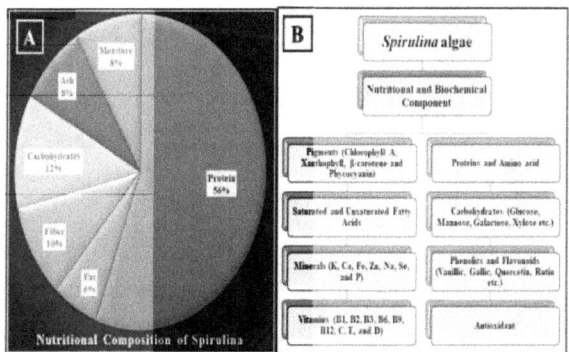

Fig. 4: A. Nutritional composition and B. biochemical components of spirulina (Source Sahara and Jood 2017).

Physical Properties of Spirulina

Spirulina has a pH of about 6.93 which may help to counter the acidic nature of certain foods. The data in Table 1 reveals the bulk density of spirulina as 0.84 Kg/lit, which is usually affected by the particle size distribution, type of agglomeration, particle porosity, and moisture content. The particle size distribution on the other hand is affected by the initial size of the filaments fed the dryer and the extent of the grinding or the pore diameter of the atomizer. The bulk density of the final product is therefore dependent on culturing, harvesting, and drying conditions. In powder form, the color of spirulina is usually blue-green to green, while it has the odor of mild seaweed.

Table 1: The physical properties of spirulina

Physical properties	Values
pH	6.93 ± 0.12
Bulk density	0.84 ± 0.02 Kg/lit
Particle size	100% 60 mesh
Appearance	Fine, uniform powder
Color	Blue green to green
Odor and taste	Mild like sea weed
Consistency	Powder

(Source: Morsy et al., 2014)

Proximate Composition of Spirulina

Proximate composition analysis of samples of commercial spirulina powder has reported 3 – 6 percent moisture, dry matter content of 94 - 97 percent, up to 60 percent protein, 20 percent carbohydrate, 5 percent fat, and 7 percent ash, thus making it a low-fat, and low-calorie source of protein. The proximate composition results of spirulina samples from different countries on a dry matter basis are shown in Table 2. Spirulina also contains numerous antioxidants such as beta-carotene, phycocyanin, tocopherols, micro-nutrients, and polyunsaturated fatty acids, like gamma-linolenic acid and phenolic compounds. The cell wall is thin, and flexible and therefore allows for easy digestion and absorption of nutrients. Spirulina is readily digested in the human and animal guts because of the absence of cellulose in its cell walls, such that after 18 hours, more than 85 percent of its protein has been digested and assimilated.

Table 2: Proximate composition results of spirulina on a percentage dry matter basis

Component	FOI, France	SAC, Thailand	IPGSR, Malaysia	BAU, Bangladesh
Crude protein	65	55–70	61	60
Soluble carbohydrate	19		14	
Crude lipid	4	5–7	6	7
Crude fiber	3	5–7		
Ash	3	3–6	9	11
Moisture		4–6	6	9
Nitrogen free extract (NFE)		15–20	4	17

FOI = French Oil Institute; SAC = Siam Algae Co. Ltd; IPGSR = Institute of Post-graduate Studies and Research laboratory, University of Malaya; BAU = Bangladesh Agricultural University (Source: FAO, 2008)

The protein content of spirulina

The major proteins are the phycocyanins, which are pigment-protein complexes that also exhibit antioxidant, and anti-inflammatory properties, and are responsible for the blue-green color of the algae. The protein complex comprises C-phycocyanin, which is known to exhibit anti-inflammatory and neuroprotective effects, and allophycocyanin which exhibits photosynthetic and antioxidant properties. Others are spirulina globulin which has antiviral and anti-inflammatory properties, spirulina albumin with antioxidant and immune-boosting properties, and phycoerythrin, a red pigment-protein complex that also exhibits photosynthetic and antioxidant properties. The algae are particularly high in the essential amino acids, lysine, methionine, and cysteine, which are often limited in plant-based diets. It is also a good source of other non-essential amino acids, such as alanine, glycine, and proline, which play important roles in the body's metabolism and overall health. The limiting amino acids in spirulina are methionine and cystine, although the values are still higher than the values obtained in grains, seeds, vegetables, and legumes. The lysine content is also higher than the values of all vegetables, except legumes. Spirulina also contains 2.2 - 3.5 percent of RNA and 0.6 - 1 percent of DNA, representing less than 5 percent of the amino acids, based on dry weight.

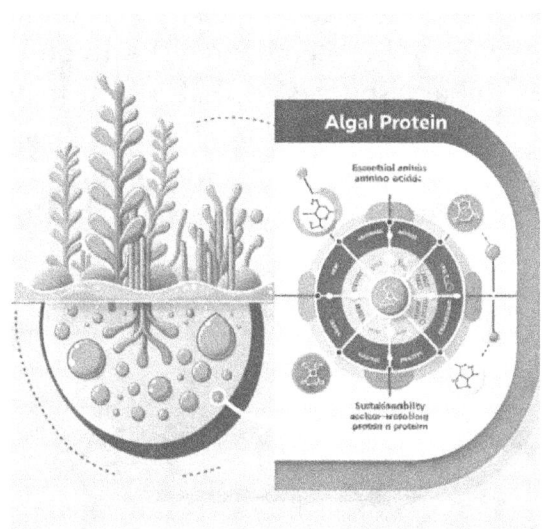

Components of algal protein (Image source: ETprotein.com)

Table 3 shows the amino acid and fatty acid values of spirulina reported by Morsy and colleagues. Table 4 compared the amount of protein found in various dietary sources to that which may be found in spirulina algae. The amount of protein in the algae fluctuate between 10 and 15 percent, depending on the time of harvest, with the most protein being present in the algae during the early morning.

Table 3: The amino acids and fatty acid contents of spirulina (mg/100 g).

Amino acids	Values	Fatty acids	Values
Essential amino acids	**%**	Myristic (C14:0)	0.52
Isoleucine	6.78	Palmitic (C16:0)	39.87
Leucine	7.67	Palmitoleic (C16:1 omega-6)	6.46
Lysine	4.37	Stearic (C18:0)	1.97
Methionine	2.39	Oleic (C18:1 omega-6)	1.83
Phenylalanine	4.42	Linoleic (C18:2 omega-6)	17.91
Threonine	4.88	Gamma-linolenic (C18:3 omega-6)	24.58
Tryptophan	1.93	Alpha-linolenic (C18:3 omega-3)	traces
Valine	6.37	Erucic acid (C22:1)	5.39
Total	38.81	Lignoceric acid (C24:0)	1.47
Non-essential amino acids	**%**	Total saturate fatty acid	43.83
Alanine	7.52	Total unsaturated fatty acid	56.17
Arginine	7.65		
Aspartic	11.17		
Cysteine	1.28		
Glutamic	13.79		
Glycine	5.24		
Histidine	2.71		
Proline	4.35		
Serine	4.16		
Tyrosin	3.48		
Total	61.19		
Total amino acids	100 %		

(Source: Morsy et al., 2014)

This amount is however significantly higher than the percentages found in animal meat and fish (15–25 percent), soybeans (35 percent), powdered milk (35 percent), peanuts (25 percent), eggs (12 percent), cereals (8–14 percent), and whole milk (3 percent). Again, according to the findings of El-Chaghaby and colleagues, the protein content of *S. platensis* at 53.30 percent (dry weight), is much higher than that of other algae species like *Chlorella vulgaris* (20.67 percent) and *Scenedesmus obliquus* (31.07 percent). Jung and coworkers reported that the plant proteins known as phycobiliprotein, which contain C-phycocyanin and allophycocyanin in a ratio of 10:1, are responsible for most of the spirulina's beneficial effects on human health. C-phycocyanin is, however, one of the primary proteins also found in moss, and accounts for about 20 percent of the total protein on a dry weight basis. This pigment is also by its molecular structure similar to biliverdin since it contains phycocyanobilin.

Table 4: A comparison of the relative protein content of spirulina algae with other food and food products

Food and Food Products			RPC * (%)
Spirulina			55-70
			55-70
			30-55
Beef			20.71
Chicken			21.96
			22.25%
Fish	Carp		16.70
	Cod		17.40
	Herring		18.10
	Salmon		18.40
Whole egg			12.60
Sausage			14.43
Milk	Buffalo		4.17
	Camel		3.38
	Cow		3.56
	Goat		3.44
	Sheep		4.35
Whey Protein	Buffalo		0.72
	Camel		0.58
	Cow		0.53
	Goat		0.54
	Sheep		0.74
Whey Proteins			54.8 3.0-74.8 4.1%
Whey Protein Concentrate Powder			33.30
White Cheese			16.20
Organic hard cheese			21.53-25.70
Soy bean			38.30-40.30
			35.35-39.80
Common Oat			11.61
Oat grains			9.70
Black Bean (Organically produced)			25.20
Maize			12.65-12.45
Rice			7.76-8.31
Wheat			11.88

(Source: AlFadhly et al., 2022)

The lipid content of spirulina

On a weight basis, spirulina is made up of 7 - 10 percent fat, which is primarily polyunsaturated fatty acids (PUFAs), such as gamma-linolenic acid (GLA), alpha-linolenic acid (ALA), and linoleic acid (LA). It also contains some saturated fatty acids, such as palmitic acid and stearic acid, and monounsaturated fatty acids, such as oleic acid. Other fatty acids found in the spirulina include stearidonic acid (SDA), eicosapentaenoic acid (EPA), docosahexaenoic acid (DHA) and arachidonic acid (AA). Tanticharoen and coworkers have shown that the PUFA, and especially the gamma-linolenic acid content of spirulina could be enhanced by cultivation under light-dark cycles in the laboratory or outdoors. The concentrations of the major fatty acids in spirulina published by Morsy and colleagues are shown in Table 3. The total saturated and unsaturated fatty acids were 43.83 and 56.17 mg/100g respectively, while palmitic acid (39.87 mg/100g), gamma-linolenic acid (24.58 mg/100g) and linoleic acid (17.91 mg/100g) were the most abundant fatty acids. Again, spirulina has been shown to have a cholesterol content of 32.5 mg/100 g; indicating that 10 g of spirulina powder will provide only 1.3 mg of cholesterol and 36 kcal of energy, while its equivalent of egg powder will provide 300 mg of cholesterol and 80 kcal of energy.

The carbohydrate content of spirulina
The carbohydrate content of spirulina is influenced by factors such as strain, cultivation, and processing methods and may range from 10 - 15 percent. The bulk of the carbohydrates is usually in the form of complex polysaccharides, while some factions of simple sugars such as glucose, fructose, and sucrose may also be present. Barrón and colleagues reported that *S. platensis* contains about 13.5 percent carbohydrates, while the sugar is mainly composed of glucose, along with rhamnose, mannose, xylose, galactose, and two unusual sugars: 2-O-mehtyl-l-rhamnose and 3-O methyl-l-rhamnose. Barrón and coworkers attributed the antiviral activity of spirulina to its content of sulfated polysaccharides, sulfoglycolipids, and a protein-bound pigment, allophycocyanin.

Vitamin and mineral content of spirulina
Spirulina contains large amounts of natural β-carotene and this β-carotene is converted into vitamin A. It is therefore particularly rich in vitamin A (in the form of beta-carotene). Ten grams of spirulina powder has been shown to provide about 23,000 IU of beta carotene. It is also vitamin B-1 (thiamin), vitamin B-2 (riboflavin), vitamin B-3 (niacin), vitamin B-6 (pyridoxine), and vitamin B-9 (folic acid). Ten grams of spirulina powder has been shown to contain 20 mcg of vitamin B-12. In addition to β-carotene, spirulina also contains several other pigments such as chlorophyll a, xanthophyll, echinenone, myxoxanthophyll,

zeaxanthin, canthaxanthin, diatoxanthin, 3-hydroxyechinenone, beta-cryptoxanthin, oscillaxanthin, plus the phycobiliproteins c-phycocyanin and allophycocyanin. The vitamin and mineral content of spirulina in mg/100g dry weight published by Koru are shown in Table 5. The mineral composition of the algae is influenced by the strain of spirulina, the location and conditions in which it is grown, and the processing method. The major minerals are phosphorus (916 mg/100g), magnesium (250 mg/100g), calcium (168 mg/100g) and iron (53.60 mg/100g), while potassium, sodium, zinc, manganese are also present in smaller amounts.

Table 5: The vitamin and mineral concentrations in spirulina powder

Provitamin A	213.00 mg
Thiamin (V.B$_1$)	1.92 mg
Riboflavin (V.B$_2$)	3.44 mg
Vitamin B$_6$	0.49 mg
Vitamin B$_{12}$	0.12 mg
Vitamin E	10.40 mg
Niacin	11.30 mg
Folic acid	40 µg
Panthothenic acid	0.94 mg
Inositol	76.00 mg
Minerals	
Phosphorus	916.00 mg
Iron	53.60 mg
Calcium	168 mg
Potassium	1.83 g
Sodium	1.09 g
Magnesium	250 mg

Anti-nutrient composition of spirulina

Although spirulina is generally a nutritionally safe algae, it may contain low levels of certain antinutrients which may interfere with digestion and nutrient absorption. For example, phycocyanin and oxalic acid found in spirulina have been shown to inhibit the absorption of iron, zinc, and calcium, by binding to them and forming complexes that are difficult for the body to absorb. It also contains small amounts of lectins and trypsin inhibitors, which can interfere with protein absorption and other nutrients. A study conducted at the King Saud University, Saudi Arabia on the concentrations of six typical heavy metals/minerals (Ni, Zn, Hg, Pt, Mg, and Mn) in 25 commercial edible spirulina products however concluded that the concentrations of inorganic elements did not exceed the regulation levels, and are therefore considered as safe food.

Conclusion

Spirulina is generally rich in proteins, vitamins, minerals, and fatty acids, but low in carbohydrate, with most of its composition being proteins and fats. It is high in the essential amino acids, lysine, methionine, and cysteine, and other non-essential amino acids, such as alanine, glycine, and proline, while the limiting amino acids are methionine and cysteine. The fat is primarily polyunsaturated fatty acids, especially the gamma-linolenic acid, while the cholesterol content is relatively low compared to that of egg. Spirulina is particularly rich in vitamin A in the form of beta-carotene, and the B vitamins such as thiamine, riboflavin, niacin, pyridoxine, and folic acid, in addition to major minerals, and several pigments that exhibit antioxidant activities.

THE FOOD VALUE OF SPIRULINA

Spirulina papaya protein bowl (Image source: foodlifestylefacts)

Prof Ifeanyi Charles Okoli

Introduction

Spirulina is considered a superfood because it is packed with nutrients, including protein, vitamins, minerals, and antioxidants. Its cell wall consists of polysaccharides which have a digestibility of 86 percent, and could be easily absorbed by the human body. There are different categories of spirulina foods, while pills and capsules made from dry spirulina are also important food supplements. It has been added to common foods such as soups, sauces, pasta, energy bars and snacks, baked goods, smoothies and juices, salad dressing, smoothie bowls, and condiments as nutritional supplements. Studies at Earthrise Farms established that more than 100 percent of the daily essential amino acid requirements for a typical adult male could be supplied by only 36 grams of spirulina. Therefore, spirulina could complement and increase the amino acid quality of vegetable protein foods. Furthermore, spirulina offers a convenient solution to the pH problems of most diets as it is very alkaline. Spirulina powder has also been used as an ingredient in an orange-flavored chewable wafer, and other types of candy, and in protein flours in which 10 percent spirulina was added to soybean or milk-egg powders, and in Pastalina, a green soy-whole wheat noodle. The enrichment of fermented foods such as cheese, yogurt, and tofu with spirulina equally offers novel opportunities for the use of spirulina in the food industry. Furthermore, the use of novel extraction methods to produce decolored spirulina powder which is odorless and tasteless, and thus suitable for widespread use has been researched.

In Chad and perhaps other neighboring African countries, spirulina is routinely used in several ways to enrich the nutrient content of local drinks, and foods such as soups and stews, where it serves as a nutrient-rich thickener. Dried spirulina cakes mixed with spices and flavorings are also eaten as a snack, while in breads, muffins, and other baked foods, it helps to boost the nutritional content and esthetic appeal of the products.

Masten Rutar and colleagues have however emphasized the need for appropriate oversight of the spirulina supplement industry, because of increasing incidents of adulteration, misbranding, and undeclared ingredients together with misleading claims that create potential risks to consumers. For example, Al-Dhabi reported the presence of toxic metals at levels of health concern in some commercial spirulina supplements, while another study by Wu and colleagues, reported the adulteration of these products with inferior cheaper materials/ingredients, such as wheat flour and mung bean powder. These practices are probably encouraged by poor quality regulation, and increased market demand as well as the high cost and complexity of producing spirulina products. Overall, spirulina is a versatile ingredient used in different foods to boost their nutritional content. Its high protein content makes it particularly valuable in areas where access to other protein sources is limited.

Improvement of the Nutritional Value of Common Foods

Extruded foods: Extruded foods have become common in the diets of a large proportion of the population of developed counties in recent times. They are usually made with ingredients or components that provide unique functional properties, although usually low in nutrient values. Several studies have shown that fortifying extuded foods with spirulina results in an increase in the amount of amino acids in treated foods since the algae is a rich source of alanine, arginine, aspartic acid, glutamic acid, leucine, and valine. Spirulina is currently integrated into a variety of goods, including baked desserts, beer, morning cereals, confectioneries, corn chips, crackers, doughnuts, food bars, frozen desserts, juice smoothies, muffins, pasta, popcorn, salad dressing, snack foods, and soups.

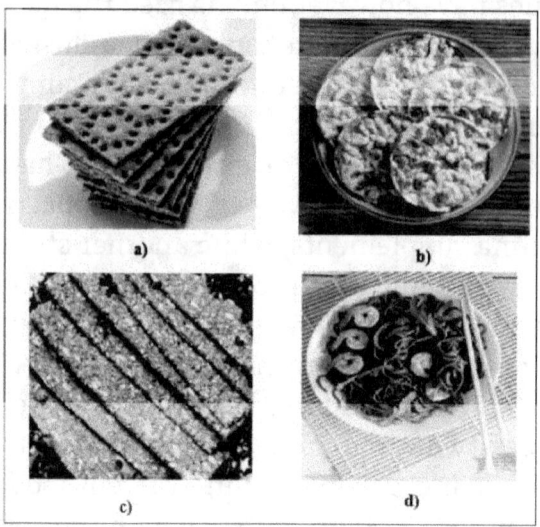

Spirulina extruded products (a) spirulina biscuits (b) spirulina rice cake
(c) spirulina energy bars (d) spirulina noodles (Image source: Jagdale, 2021)

There are numerous cookbooks on how to apply spirulina in these food preparations. Gautam and colleagues reported that the nutritional profile of handmade extruded goods manufactured from a composite of foxtail millet, wheat, chickpea flour and spirulina powder has a great deal of potential for meeting the nutritional needs of vegetarians, and the impoverished in society. Among the Kanembu people d*ihé* (spirulina cake) is crumbled and mixed with a sauce of tomatoes and peppers, poured over millet, beans, fish, or meat, and in 70 percent of their meals. Pregnant women eat *dihé* cakes directly because they believe their dark color will screen their unborn baby from the eyes of sorcerers.

Fatima and Srivastava evaluated the organoleptic, and storage characteristics of some food items fortified with spirulina at 10 and 5 percent inclusion rates, and reported that the food products are healthy, and have a considerably higher nutritional value than control samples. The product has also higher extrusion potential and a greater acceptance based on organoleptic criteria, implying their improved quality and benefit to consumers. A value-added extruded product containing 5 percent spirulina, 95 percent wheat flour, and 5 percent corn flour was also produced by Lakshmi and coworkers, and the sensory criteria such as taste, odor, texture, and color were found acceptable. In another Indian study, recipes supplemented with spray-dried spirulina powder at 1, 2.5, and 5 g were ranked according to the degree of acceptance. They reported that all the recipes incorporated with spirulina were acceptable in appearance/color, texture, taste/flavor, and overall acceptability at 1 g and 2.5 g levels. They concluded that spray-dried spirulina could be effectively incorporated into various Indian recipes to improve their nutritional qualities, and application in solving various health and dietary challenges.

Dairy products: Functional dairy products such as milk, yogurt, and cheese have also been supplemented with spirulina algae. For example, yogurt has been made using dried spirulina at 0.1, 0.2, 0.3, and 0.5 percent concentrations, with the 0.3 percent concentration giving the best positive effect on nutritional and sensory value enhancement. Spirulina in yogurt also improves the survivability of *S. thermophilus* and *L. bulgaricus* during storage. Guldas and colleagues included up to one percent of spirulina during the preparation of yogurt, and recorded a positive effect on the number of lactic acid bacteria, while the addition of 0.5 percent was superior to the one percent in terms of the sensory characteristics of the products. In an earlier study, Gouveia and coworkers added *S. platensis* to milk that had been fermented by probiotic bacteria using a starter that contained *L. acidophilus, Bifidobacteria*, and *Streptococcus thermophilus.* They reported a beneficial impact in terms of increasing the number of initiator bacteria that survived during the storage period.

The inclusion of spirulina in ice cream has also been tried due to its content of antioxidants, such as polycarotene phenols, which have the ability to scavenge free radicals thereby, having positive impacts on health. For example, Szmejda and colleagues added spirulina extract to ice cream, and reported an increase in antioxidant activity, with the highest inhibition level of 39.7 percent in mint ice cream samples, while the control value was 32.8 percent. Spirulina has also been employed as a stabilizer in the production of ice cream at replacement levels of 25, 50, 75, and 100 percent, although the optimal replacement level was found to be 50 percent at a concentration of 0.15 percent. One percent inclusion of *S. platensis* during the preparation of a soft cheese product was also found to be the optimal concentration from both the physicochemical and the sensory points of view, since it had a favorable influence on the protein, water, fat, and β-carotene levels of the final product.

Traditional foods: Spirulina has also been used to prepare traditional foods such as instant noodles, stylish noodles, nutritious blocks, beverages, and cookies as luncheon food for middle-grade students. For example, the inclusion of spirulina in the production of dried noodles has been shown to improve the nutritional quality of the noddles because of the high protein and β-carotene content of the spirulina. In a study at the Diponegoro University, Indonesia, researchers determined the effects of 7, 9, and 11 percent inclusion of *S. platensis* paste on the physical, chemical, and sensory characteristics of dried noddles. The study showed that a 9 percent inclusion of spirulina paste resulted in a significant effect on the elasticity, β-carotene, water, protein, ash, fat, and carbohydrate contents, and the sensory characteristics of the dried noddles, compared to the control, and the other inclusion levels. They therefore concluded that a 9 percent inclusion of *S. platensis* paste could significantly improve the nutritional quality of dried noodles, especially the protein and β-carotene contents.

Effect on malnutrition: Studies have also shown that as little as 10 g a day of spirulina could result in rapid recovery from malnutrition, especially in infants. For example, in Togo, rapid recovery of malnourished infants was reported in a remote village health clinic, where children were given 10 to 15 g spirulina per day as a dietary supplement in a millet and spices-based diet and they recovered in several weeks. Again, a report from China, showed that prescription of spirulina at Nanjing Children's Hospital as part of a "baby nourishing formula" with baked barley sprouts resulted in 27 out of 30 children aged 2 to 5 years recovering in a short period from bad appetite, night sweat, diarrhea, and constipation.

Spirulina in Vegetarian and Vegan Diets

Recent innovative research findings have highlighted the importance of functional vegetarian diets in improving human health, and minimizing disease risks. Practically, adopting a vegetarian dietary pattern is traditionally interpreted to mean an absence of meat products, and high consumption of fruit, vegetables, legumes, nuts, grains, and soy protein food components which has been associated with positive health impacts. Vegetarian diets have been classified into lacto-ovo-vegetarians (includes dairy and egg), lacto-vegetarians (includes dairy), ovo-vegetarians (includes egg), and vegan which excludes all animal-origin foods. Spirulina has been proven to be a suitable supplement for vegetarian and vegan diets since it contains up to 60 percent protein and essential amino acids not matched by any other plant food. While many plant-based protein sources such as legumes and grains require combinations with other foods to ensure adequate amino acid balance, spirulina's nutritional profile meets this quest to balance the vegetarian diet.

(Spirulina power cookies. Image source: Jenny's blog Healthy Crush)

Spirulina nutritional profile also allows for easy incorporation into diets without significant changes in taste and texture. Recent studies have shown that with the increasing global population of people who identify as vegetarians or vegans, which is about 14 percent in 2021, there is the possibility that spirulina will be increasingly consumed by a much larger percentage of this population than the traditional whey or casein-based protein powders. A recent study by Atik and colleagues assessed the value of *S. platensis* fortification in vegan kefir containing soy and almond at 0.25 and 0.5 percent. They reported that increasing the spirulina concentration increased the counts of lactobacilli and lactococci, and the total phenolic content of the kefir, while the pH decreased. They concluded that spirulina is a promising functional food component for improving the probiotic potential of bioactive vegan food. In an earlier study by Ghazal, three innovative ready-to-use (RTU) and ready-to-eat (RTE) ovo-vegetarian diets (UVD) incorporating different vegetables (pea, taro, and broccoli) at 15 percent in chicken pea-based diet was prepared with or without one percent spirulina inclusion and analyzed for their nutritional and bioactive properties after cooking. Composite analysis results were 62.46 to 68.54, 17.52 to 20.57, 5.54 to 6.19, 6.97 to 8.92, 5.09 to 6.65, and 61.49 to 63.84 percent for moisture, crude protein, lipids, ash, fiber and available carbohydrate contents in the RTU OVDs, respectively, indicating significant nutritional improvements. Higher organoleptic acceptability of the RTE OVDs was also observed, confirming their attractiveness to consumers. The study concluded that the production of RTU and RTE OVDs incorporated with commonly consumed vegetables, and supplemented with spirulina is a promising approach to improving ovo-vegetarian diets, and thus healthy human dietary practices.

Spirulina in Sports Nutrition Drinks

The higher relative usage of stored fat during exercise implies enhanced preservation of glycogen stores, which could be rate-limiting for endurance during prolonged moderate-intensity exertion. The universal isotonic drinks used in competitive sports usually contain functional ingredients, such as proteins for the formation and maintenance of muscle mass, or bioflavonoids as antioxidants that prevent oxidative processes. According to Sanchez-Oliver and coworkers, in creating protein-containing sports drinks, choosing a protein ingredient is determined primarily by the nutritional value and degree of assimilation of the protein source. Manufacturers usually prefer whey protein (48 percent), casein (22 percent), whey-casein protein (18 percent), soy protein (8 percent), and egg protein (4 percent). Some studies have shown that *S. platensis* could serve as an alternative protein source since the protein content of the dry powder is up to 60 percent. It is also rich in omega 3 and 6 polyunsaturated fatty acids. Zeinnalian and colleagues specifically reported that the use of spirulina during exercise increases muscle endurance by 20 - 30 percent. A study conducted by Kalafati and colleagues investigated the impact of 6 g daily supplementation of spirulina on exercise. The subjects exercised at sub-maximal intensity on a treadmill for 2 hours and immediately after, sprinted at maximal intensity on the treadmill until they became exhausted. They observed that the time until exhaustion was more than 30 percent longer among the subjects who had been taking spirulina. In addition, spirulina supplementation was able to prevent exercise-induced increases in blood-based, oxidative-stress markers and also increased the net contribution of fat-burning to the submaximal exercise, and was attributed to the protection of muscle mitochondria from oxidative stress.

Researchers at the Siberian Federal University, Krasnoyarsk, Russia, evaluated the antioxidant activity (AOA) of an aqueous extract of spirulina by *in vitro* auto-oxidation of adrenaline in an alkaline medium, to determine the biological value of spirulina proteins. Since the athletes diets is usually characterized by a lack of protein of plant origin, vitamins, and antioxidants, the researchers selected spirulina, fruits, and wild berries as sources of biologically active substances. The technological scheme for preparing the sports is shown in Figure 4. The quantitative content of spirulina in the protein fruit and berry drinks was determined using comprehensive quality assessment by organoleptic indicators. They reported that the water extracts of spirulina powder exhibited antioxidant activity due to the presence polyphenols, bioflavonoids, phycocyanin, and chlorophyll pigments. They reported that using a comprehensive quality assessment, the spirulina dosage in the smoothie is justified. They also noted the positive effect of the developed drink containing protein and a vitamin-mineral complex on competitive activity. They therefore recommended the spirulina sports drinks for industrial production as a means of promoting healthy nutrition, and attracting young people to sports nutrition without prohibited drugs.

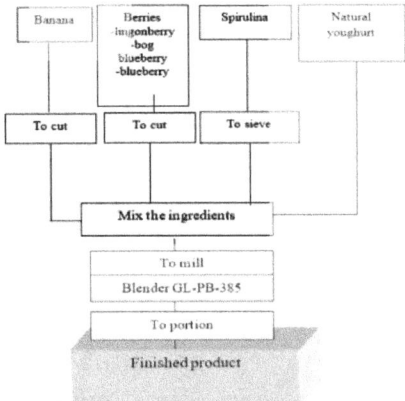

Fig. 4: The technological scheme for preparing a drink for sports nutrition (Source: Aleksandrovna et al., 2019)

Consumer Acceptance of Spirulina

The organoleptic properties such as appearance (color and porosity), taste (bran flavor, bitterness, off-odor, and aftertaste), texture (hardness, crispness, brittleness, and firmness), odor (odor raw material, stink odor, undesirable odor, and old odor), solubility in mouth, volume, and overall acceptability are the major factors that govern consumer acceptance of a food product. Mosy and colleagues used the sensory evaluation method to select the best ratio of addition of spirulina in snacks. Sensory properties were considered one of the limiting factors of consumer acceptability, with the data indicating that significant changes were recorded in all the organoleptic properties of all experimental products. In many countries, especially in the West, spirulina is used mostly as a food supplement, additive, or dye, and is marketed as powder, pills, or capsules, while its actual use as a food is less widespread. Researchers at the University of Gottingen, Germany, in a recent study, developed novel spirulina-based food products to promote its broader use as a food ingredient. The three innovative products: pasta filled with spirulina, maki-sushi filled with spirulina, and spirulina jerky were offered to consumers willing to taste and evaluate them. Overall, the spirulina-filled pasta was the most preferred product, partly due to familiarity with products in that category more than sushi and jerky. The researchers however concluded that all the spirulina product concepts would work equally well, if pasta, sushi, and jerky were similarly familiar to the target consumer population. This is because all the tested benefits were equally accepted with each product, except that spirulina jerky would have to be marketed as an innovative product.

Nutritional Safety of Spirulina Products

After cultivation, the commercialization of the spirulina algae requires that it be processed, packaged, and distributed to the market and the consumers. These processes could lead to substantial changes in the chemical composition of the algae, thus affecting its nutritional quality and toxicological properties. For example, a 2013 study by Al-Dhabi, highlighted the presence of anomalous quantities of cadmium, mercury, arsenic, and lead in spirulina products, and attributed this to pesticide or fertilizer use adjacent to the spirulina cultivation sites. Tidjani and colleagues also isolated higher bacterial counts (25 and 3.7×10^5 CFU/g) than reference values (3×10^5 CFU/g) from a locally produced spirulina cake (Dihe) sample sold in a local market in Chad, while another sample contained some unidentified molds. The study therefore shows that locally produced spirulina cakes could be contaminated by potentially harmful microorganisms. Such contamination is of concern because, in some countries, dietary supplements may not be strictly regulated, or inspected to the same extent as other food and pharmaceutical products.

Spirulina species should be properly identified before cultivation

(Image source: www.cfshungary.hu)

Since different types of spirulina species grow at different ecological sites, it is important to properly identify the species being cultured either for personal consumption or for commercial purposes. Microcystins and hepatotoxins, producing spirulina species may cause acute poisoning, liver damage, gastrointestinal disturbances, and cancer. Some studies have associated long-term consumption of spirulina with the pathogenesis of Alzheimer's and Parkinson's diseases. Lu and colleagues have also reported other side effects linked to the ingestion of spirulina such as headache, stomach ache, muscle pain, sweating, difficulty in concentration, and rapid dissolution of damaged skeletal muscles characterized by muscular weakness), swelling, cramping, and dark-colored urine. Again, the presence of high levels of the amino acid, phenylalanine in spirulina may be a major disadvantage for people suffering from phenylketonuria, caused by an absence or deficiency of an enzyme called phenylalanine hydroxylase. The high nucleic acid content in spirulina could equally increase the levels of uric acid in the blood leading to gout, while differences in the levels of blood glutathione peroxidase, and lactate dehydrogenase that catalyzes pyruvate metabolism have also been associated with spirulina supplementation.

About 80 percent of the spirulina products in the market are made from *S. platensis*, which has also received tremendous research attention, and may therefore be deemed the appropriate species to be cultured for human consumption. Gogna and colleagues reported that the safe recommended dosage of spirulina for adults is approximately 3 -10 g/day, with 30 g/day being the maximum limit for consumption.

Quality Control of Spirulina Products

The increasing market demand, high cost, and complexity of culturing spirulina may encourage the adulteration of its products with inferior, and cheaper ingredients, such as wheat flour and mung bean powder, resulting in economic losses, and potential health risks to consumers. Researchers at the Jožef Stefan Institute, Ljubljana, Slovenia characterized the elemental, amino acid, and fatty acid content of commercially available spirulina supplements in Slovenia and compared the results with their nutritional declarations. The data generated confirm that the spirulina supplements were good sources of calcium, phosphorous, potassium, and selenium, while the iron content although relatively high, may not be readily bio-available since it is mainly present as the ferric cation. More importantly, the study revealed that the presence of additives in the spirulina supplements resulted in significant variations in their nutrient content, and in some instances lower product quality. Again, a very high 86.7 percent of the product declarations were inconsistent with the elemental content results obtained from the study. The study therefore highlights the need for a stricter control system for spirulina-based supplements sold in the market. Other USA and Canadian studies have reported the absence or minimally below the regulatory levels of microcystins and hepatotoxins in commercial spirulina products sold in those countries. Combating the contamination and adulteration of spirulina products requires proper regulatory frameworks which will help to improve consumer trust, and guarantee the availability of quality, and safe spirulina products in the market.

Conclusion

The spirulina algae are both biologically and economically important due to the numerous applications that have been developed for them in the food industry. The products are mostly available as food supplements, additives, or colorants, and are mostly marketed as powders, pills, or capsules, while the actual use of spirulina as a food is less widespread. There is a need for proper regulation of commercial spirulina products to combat cases of contamination and adulteration, and to ensure consumer trust and availability of quality and safe spirulina products in the market.

THE HEALTH BENEFITS OF SPIRULINA ALGAE-1

(Spirulina powder and pills. Image source, Dr Axe)

Introduction

The changing lifestyle, and food habits that have resulted in the current consumer quest for organic foods, and nutraceutical products has heightened the global research attention on natural food supplements like spirulina which are rich in phyto-nutrients, and bioactive metabolites. Lyons and colleagues observed that the adverse side effects commonly associated with several drug therapies have led to growing public interest in natural products with health-promoting properties as an alternative to conventional drugs. Beginning in the 1980s, extensive investigations have therefore been devoted to the development of functional foods for preventing or managing various diseases. Alga are the oldest oxygenic photosynthetic organisms known so far, and they also serve as a rich source of novel bioactive metabolites, including many cytotoxic, anti-fungal, and antiviral compounds. Spirulina algae is indeed believed to be one of the most important healing, and prophylactic nutritional supplements of the 21st century due to its nutrient profile, therapeutic effects, and low toxicity. The consumption of spirulina has been shown to have several potential health benefits because of its rich nutrient and antioxidant contents. It has however been reported that *S. platensis* is the most commercially available spirulina food supplement, representing about 81.2 – 100.0 percent of the commercially available alga products.

The phytonutrient and pigment contents of the algae are increasingly receiving research attention because of their proven physiological and therapeutic effects on oxidative injury, and their nutraceutical and potential pharmaceutical properties. Numerous studies have indeed documented the high nutritional value, eco-friendly, and curative capabilities of spirulina. It has also been proven that spirulina could play many roles in the mitigation of many health-compromising conditions such as reduction of blood sugar levels, the reduction of blood pressure, and the modification of dysbiosis caused by an imbalance in the intestinal flora. Several other studies have also confirmed the anticancer, anti-diabetic, anti-inflammatory, immunomodulatory, anti-aging, anti-allergic, anti-anemic, nephroprotective, neuroprotective, and hepatoprotective effects of spirulina on both human and animal subjects as highlighteed in figure 5.

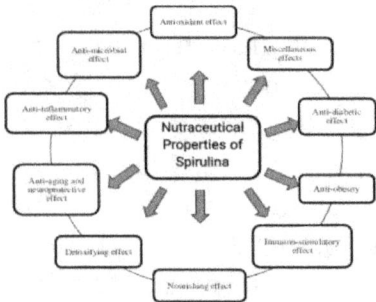

Fig. 5: The nutraceutical properties of spirulina (Source: Priyanka et al., 2023)

Other studies have also highlighted the beneficial effects of spirulina on heart diseases in humans, and the reproductive performance of animals. Spirulina capsules have proved effective in lowering blood lipid levels, and in decreasing the number of white blood cells associated with radiotherapy and chemotherapy. These health benefits have been attributed primarily to the chemical constituents of the algae which include minerals (especially Fe), phenols, phycocyanin, and polysaccharides. Siva Kiran and colleagues have therefore proposed the extensive global cultivation and consumption of spirulina as a sustainable approach to preventing Protein Energy Malnutrition (PEM) and Protein Energy Wasting (PEW) in humans.

Spirulina as a Nutritional Supplement

Spirulina is rich in quality protein, vitamins, minerals, and polysaccharides which are highly digestible, and easily absorbed by the human body. The polysaccharides are characterized by their high molecular weight, are composed of multiple monosaccharides connected by glycosidic bonds, and have been shown to have substantial biological activity. Gut microbiota can break down the spirulina polysaccharides in to butyrate, short-chain fatty acids, and other metabolites that can readily be absorbed, and utilized by the body.

Different brands of spirulina pills and capsules have been made from dry spirulina powder. Chinese studies show that female athletes recorded increases in their haemochrome level after taking 10 g of spirulina pills per day for four weeks, while there was no apparent increase in the male athletes. The lung capacity of juvenile weight-lifters and *Jujutsu* athletes was also improved. In Vietnam, *S. platensis* powder health food tablets are marketed with the brand names Linavina and Pirulamin, while a canned product called Lactogil is also promoted as an enhancer of milk secretion in mothers having poor lactation. Spirulina powder is also being used in hospitals to treat highly malnourished children in that country. Again, the high concentrations of essential PUFAs such as gamma-linolenic acid (GLA), in spirulina have been exploited in the treatment of many degenerative diseases. Early studies by Henrikson reported that 10 g of spirulina supplement, which contains more than 100 mg of GLA was beneficial in treating arthritis, heart disease, obesity, and zinc deficiency. The very high iron content of spirulina has also been exploited in the treatment of anemic conditions. For example, a Japanese study reported that eight young women presenting hypochronic anemia, and given 4 g of spirulina supplement each after meal for 30 days, recorded improvements in blood hemoglobin content from 10.9 to 13.2 indicating a 21 percent increase. An Indian study conducted on 20 males and 20 females by Anuradha and Vidhya reported significant reductions in blood glucose levels due to spirulina supplementation in their diets.

The Effect of Spirulina on Intestinal Flora Balance

The gut microbiota plays an important role in regulating metabolism, mucosal immune functions, vitamin production, and digestion, thereby aiding the stability of the gut environment. Biological, dietary, or drug factors have however been shown to disrupt the gut microbiota, leading to impaired gut functions and metabolic disorders such as diabetes, obesity, and metabolic syndrome. Several studies have reported the ability of spirulina polysaccharides to regulate gut microbiota disorder, thus, fostering general well-being, and averting the onset of diseases. The spirulina polysaccharides are primarily composed of rhamnose, xylose, glucose, galactose, and glucuronic acid, while the main structural characteristics are determined by average molecular weight, monosaccharide composition, and glycosidic bond position. Aqueous extract of *S. platensis* containing xylose, galactose, oligosaccharides, and resistant starch has been shown to boost the development of probiotic bacteria, and encourage the growth of beneficial bacteria such as lactic acid bacteria, *Bifidobacteria*, and *Lactobacilli* in the colon, while also reducing the number of the pathogenic ones like *Enterobacteria* and Clostroids.

Recent studies have therefore considered the use of *S. platensis* as a prebiotic source because it can benefit the growth of many gut microbiota such as *Akkermansia*, *Lactobacillus*, and *Butyricimonas*, while suppressing the growth of pathogens such as *Clostridium* and *Dorea in vitro*. Other studies have also shown that *S. platensis* could largely reduce the relative numbers of *Proteobacteria* and the *Firmicutes/Bacteroidetes* ratio in fecal samples of high fat-diet-fed rats. In constipation, disorders in gut microbiota and metabolites lead to abnormal intestinal movement and intestinal hormone secretion, affecting intestinal function. Guan and coworkers however reported that spirulina polysaccharide could improve the activity of xylanase and protease in the intestinal tract of constipated mice, accelerate chyme digestion, improve the intestinal micro-environment, reduce the number of harmful bacteria in the intestinal tract of animals, increase the number of beneficial bacteria in the intestinal tract of mice, and maintain the stability of gut microbiota. Figure 7 highlights the specific ways in which the consumption of spirulina algae might influence the intestinal microbial balance.

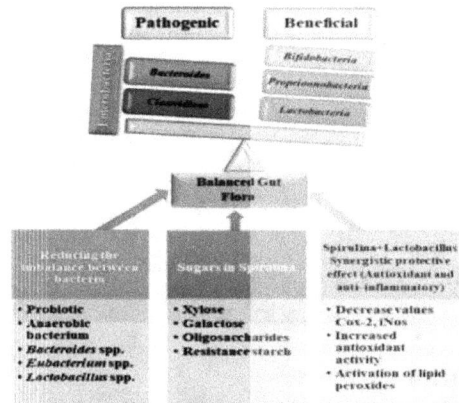

Fig. 7: The effect of spirulina on the gut flora balance (Source: AlFadhly et al., 2022)

Detoxification Effects of Spirulina

Again, because spirulina is an alkaline food, it could be used to counter the effects of acidic foods by raising the pH level of such foods, and subsequently improving digestion. Consuming more alkaline foods has been linked with improvements in immune, mental, and kidney functions, and higher energy levels, among other important benefits. Spirulina has also the property of being able to chelate toxic minerals. Thus, it can be used to detoxify arsenic and heavy metal contaminated water and food. A Chinese study specifically demonstrated the ability of bioactive extracts from spirulina to neutralize the toxic effects of heavy metals, in addition to anti-tumor activity. Several studies demonstrate the deleterious impact of repeated ingestion of high doses of monosodium glutamate (MSG) on laboratory animals and humans. Particularly in humans, repeated ingestion of high doses of MSG has been shown to trigger an elevated frequency of nausea and headaches. Researchers at the University of Agricultural Sciences and Veterinary Medicine, Calea Mănă, Romania investigated the possible toxic effects of MSG, and the ameliorating protective effects of the dietary *S. platensis* supplementation on the blood biochemical parameters, and the damage produced in organs of the Swiss mice after applying a supplementary daily dose of 0.5 and 1.0 mL of MSG with or without 1 mL of spirulina for 4 weeks. They observed an abnormal passive behavior in the MSG-treated rats with histopathological changes in their kidneys resulting in modifications to important serum biochemical parameters. The Swiss mice treated with MSG, and supplemented with spirulina showed no histopathological changes in the organs, while the serum biochemical values were within the normal range for healthy Swiss mice, indicating the ameliorative effects of spirulina on MSG toxicity.

Anti-Microbial Effect of Spirulina

Spirulina platensis can inhibit the growth of several Gram-negative and Gram-positive bacteria through the action of its extracellular metabolites. Methanol extract from spirulina culture medium has been shown to exhibit higher antimicrobial activity than other antimicrobial agents such as hexane, dichloromethane, petroleum ether, ethyl acetate extracts, and volatile components against Gram-positive and negative bacteria, and *Candida albicans*. As a natural ingredient, spirulina has also been used to treat acne and other bacterial infections. Spirulina methanolic extracts have been shown to inhibit acne-causing bacteria such as *Staphylococcus aureus* and *Pseudonomas aeruginosa*. Researchers at Bogor Agricultural University, Bogor, Indonesia synthesized spirulina extract-water soluble chitosan nanoparticles using the ionic gelation method, and measured the antibacterial activity of the extract and its nanoparticles. The yield of the ethanolic extract was 13.87±4.16 percent with the bioactive components being saponins, tannins, steroids, and phenols. The extract-loaded chitosan nanoparticles were fabricated with a nano-range size and narrow polydispersity, with the amount of loading of the extract affecting the size and dispersion of nanoparticles. They reported higher antibacterial activity by the nanoparticles than spirulina extract against *S. aureus* and *P. aeruginosa*, indicating that nano-encapsulation could enhance the extract uptake by bacteria cells, thereby increasing its potency as a cosmeceutical products. Spirulina has also been used to synthesize bio-functionalized gold nanoparticles with antibacterial activity against Gram-positive organisms such as *B. subtilis* and *S. aureus*. The abundant phenolic compounds in the spirulina biomass are responsible for antimicrobial or bacteriostatic activities.

An odorless yellowish-green compound soluble in methanol, diethyl ether, chloroform, and dimethyl sulfoxide, but sparingly soluble in water, and acetone was purified from *S. platensis* by El-Sheekh and coworkers, and found to be active against the yeast, *C. albicans*, and the Gram-positive *B. subtilis* at lower concentrations in comparison to the effect against the Gram-negative bacterium, *P. aeruginosa* at higher concentrations. During fermentation however, spirulina biomass has a stimulatory effect on the growth and/or increases the survival during storage of *Bifidobacterium*, *L. acidophilus*, *L. bulgaricus*, *L. casei*, and *S. thermophilus* which are used as starter bacteria for the production of yogurt. These positive effects on growth and survival these beneficial organisms have been attributed to the high level of free amino acids and phenolic compounds in the spirulina biomass. Figure 8 highlights the basic microbial-modulating activities of spirulina, prevention of dysbiosis, and protection of the host from infections.

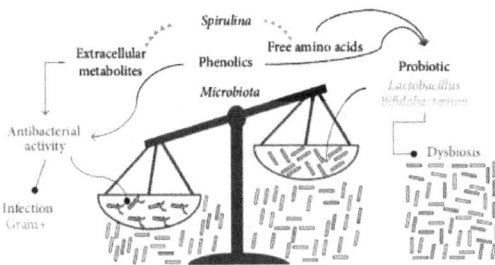

Fig. 8: The basic microbial-modulating activities of spirulina, prevent dysbiosis, and protect the host from infections (Source: Finamore et al., 2017)

The Anti-Viral Effects of Spirulina

Several studies have established the potential antiviral properties of spirulina. Simple water extracts of spirulina, and even dried spirulina biomass have been shown to possess antiviral activities. Japanese researchers were among the first to isolate a sulfated polysaccharide from spirulina that exhibited antiviral activity. These researchers demonstrated the effectiveness of the compound which they named 'Calcium Spirulan" against a variety of viruses such as *Herpes simplex* type 1, human cytomegalovirus, measles, mumps, influenza A, and HIV-1, through its ability to inhibit the replication of their virus envelopes. Calcium Spirulan also selectively inhibited the penetration of the virus into host cells. Therefore, the researchers concluded that Calcium Spirulan could be a candidate anti-HIV therapeutic drug. Another Japanese study by Teas and colleagues used existing literature to establish HIV inhibition by the algae *in vivo* and *in vitro*. There is also epidemiological evidence that populations associated with high algae consumption have correspondingly lower HIV infection. For example, they showed that within Africa the rates of HIV infection vary between different countries, and groups of people, with Chad having a very low reported rate of HIV/AIDS compared to the rest of Africa. Again, the Kanembu people of Chad who consume between 3 to 13 grams of spirulina per day had a very low reported cases of HIV/AIDS. They therefore concluded that the regular consumption of spirulina could help prevent HIV infection, and decrease viral loading of those infected, and recommended its regular consumption by this population group. Another study by researchers at the Harvard Medical School, USA reported that spirulina water extract inhibited HIV-1 due to the activities of the polysaccharide and tannin fractions of the extract. The researchers also concluded that aqueous *S. platensis* extracts have anti-retroviral activity that may be of potential clinical interest.

Other components of the spirulina algae have equally been

shown to play some roles in its antiviral activity. For example, studies at Cairo University, Giza, Egypt, reported that phosphate buffer, and water extracts of salt-stressed *S. platensis* exhibited antiviral activities against both Hepatitis-A-virus-type-MBB (HAV-MBB strain, RNA virus), and Herpes simplex-virus-type-1 (HSV-1, DNA virus) with the water extracts being more effective than phosphate buffer extracts in inducing antiviral activities, especially against HSV-1 virus. Biochemical analysis of the salt-stressed algae revealed that the lipid content was slightly increased in certain saturated and unsaturated fatty acids, especially the polyunsaturated ones (γ-linolenic acid, omega 3 fatty acid). Electrophoretic analysis of the soluble proteins showed that five new protein bands were recorded, in addition to an increase in the intensity of six existing bands. Evidence derived by Siedenburg and Cauchi from meta analyse of 25 previously published articles tended to suggest that spirulina may have prophylactic and therapeutic efficacy against SARS-CoV-2 via several pathways, though further investigation is needed to verify the linkages identified.

The Effect of Spirulina on Heavy Metal Toxicity

Spirulina has been shown to neutralize or chelate toxic minerals. For example, researchers at Beijing University, China reported that bioactive molecules extracted from spirulina could neutralize the toxic effects of heavy metals, indicating that the algae could also be used to ameliorate the toxic effects of such minerals, especially arsenic from water, food, and the environment. An earlier study successfully showed that spirulina counteracted the effects of heavy metals on the kidneys by potentiating the detoxification process. The protective effect of *S. platensis* against cadmium-induced oxidative stress, which could be either indirect or direct has been linked to the high concentration of antioxidants in the algae, and its stimulation for the synthesis of nitric oxide that supports the antioxidant system. Again, the protective role of spirulina against cadmium and lead toxicity has been attributed to the alteration in the T lymphocyte, reticulocyte, red and white blood cell counts, the mean cell hemoglobin concentration, hemoglobin, and polycythemia vera that follows spirulina supplementation. *Spirulina platensis* has also been shown to affect the iron and hemoglobin metabolism in rats subjected to lead, cadmium, zinc, and mercury-induced poisoning through its metal-binding capacity. The phycocyanin pigment in spirulina has equally been shown to enhance the activities of certain cellular antioxidant enzymes. Thus, the metal protective role of spirulina could be related to its high contents of vitamins E and C, beta-carotene, and the enzyme superoxide dismutase, selenium, and phycocyanin.

Conclusion
Spirulina algae is an important healing, and prophylactic nutritional supplement due to its nutrient profile, therapeutic effects, and low toxicity. Its proteins, vitamins, minerals, and polysaccharides are highly digestible, and easily absorbed by the human body. Spirulina polysaccharides can regulate gut microbiota disorder, boost the development of probiotic bacteria, and encourage the growth of beneficial bacteria such as lactic acid bacteria. Bioactive extracts from spirulina can neutralize the toxic effects of heavy metals and mono sodium glutamate, and also exhibit anti-microbial and anti-viral activities. The phytonutrients and pigments in spirulina have proven physiological and therapeutic effects on oxidative injury, as well as nutraceutical and pharmaceutical properties.

THE HEALTH BENEFITS OF SPIRULINA ALGAE-2

Spirulina powder scoop (Image source: PixelsAway)

Introduction

The health benefits of spirulina extend beyond its nutrient density to the complex roles played by its antioxidants in combating oxidative stress, which is linked to various chronic diseases, including cancers and heart diseases. Furthermore, spirulina has been shown to enhance immune functions, and regulate blood sugar levels, making it an attractive option for individuals seeking to bolster their overall health and well-being. Thus, researchers have explored its role in managing various health conditions, from ameliorating metabolic disorders to immune system support. Again, a notable pharmaceutical application of spirulina is its anti-inflammatory properties. Chronic inflammation is a contributory factor to numerous health issues, including heart disease, diabetes, and arthritis. Studies have demonstrated that spirulina can modulate inflammatory responses in the body through its ability to inhibit the release of histamine from mast cells, potentially offering a natural alternative to conventional anti-inflammatory medications. In addition to its anti-inflammatory effects, spirulina has shown potential to support cardiovascular health. Research indicates that spirulina consumption can lead to reductions in cholesterol levels, and improvements in lipid profiles. Its ability to enhance nitric oxide production may also aid in improving blood vessel function and circulation. These findings have led to the incorporation of spirulina into pharmaceutical formulations designed to combat cardiovascular diseases, further expanding its role within the healthcare sector.

Another significant area of spirulina research is its potential in oncology since certain compounds identified in the algae have been shown to exhibit anti-cancer properties, by inhibiting tumor growth, and enhancing the efficacy of traditional cancer treatments. Research is however ongoing to better understand these mechanisms, and to identify how spirulina can be effectively integrated into comprehensive cancer care strategies. As the demand for natural and plant-based remedies continues to rise, the pharmaceutical industry will increasingly be looking for new sources of therapeutic agents such as spirulina as viable ingredients in the formulation of new drugs and supplements. Its rich nutrient profile, safety, and sustainability, particularly make spirulina an appealing choice for researchers and manufacturers alike. The ongoing exploration of spirulina's therapeutic benefits therefore signifies a promising future for its applications in pharmaceuticals, potentially leading to innovative treatments that align with the growing consumer preference for natural health solutions.

The Role of Spirulina in Regulation of Immunity

Spirulina plays a major role in boosting immunity through its wide range of biological activities, and the high concentration of natural nutrients it contains. First, spirulina is a rich source of antioxidants and anti-inflammatory substances, which not only support and develop physiological functions of the nervous system and brain, as well as compensate for nutritional deficiencies, but also promote a beneficial immune response. Secondly, it can modulate the immune system and the bioactivity of macrophages by activating T and B cells, promoting the production of antibodies and cytokines, increasing the concentration of natural killer cells (NK cells) in tissues, and encouraging the generation of antibodies as shown in Figure 9. Spirulina is a stimulant of immune system cells because it increases anti-inflammatory resistance, stimulates the synthesis of antibodies and cytokines, activates macrophages, stimulates natural killer cells, and activates T and B cells.

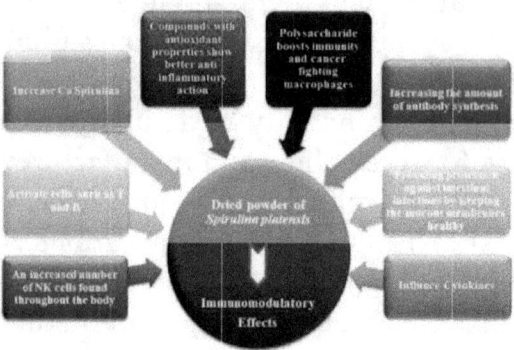

Fig. 9: Role of spirulina in the regulation of immunity (Source: AlFadhly et al., 2022)

The *S. platensis* is rich in minerals and carbohydrates consisting of polymers, including glucose and branching polysaccharides comparable to glycogen, which have been shown to be largely responsible for its positive effects on human health. These high molecular weight compounds of negatively charged sugars referred to as "Immulina," have been shown to enhance the activity of natural degrading cells against cancer cells when ingestion at a rate of 400 mg per day for seven days. The antiviral activity of these drugs as stated earlier has been demonstrated against types I and II of the herpes simplex virus. The immuno-reactivity is mostly attributed to Ca spirulan, which comprises saccharides: rhamnose, methyl rhamnose (acofriose)-3-O, methylrhamnose di-O-2,3, methylxylose-3-O, uronic acids, and sulfates. The phycobiliproteins produced from spirulina are also utilized in clinical medicine, immunological analysis, therapeutic and diagnostic cases as fluorescent material. Its ability to considerably lower the blood cholesterol level also protects against hepatitis. In the gut, spirulina aids intestinal epithelial integrity, the first line of defense of the mucosal barrier against infections.

The Role of Spirulina in the Management of Obesity

Current literature supports the use of spirulina in reducing body fat, waist circumference, body mass index, and appetite through its effects on blood lipids. Spirulina has been hypothesized to regulate body weight by inhibiting the migration of macrophages into visceral fat, thus preventing fat build-up in the liver, lowering the levels of oxidative stress in the body, and increasing insulin sensitivity and satiety as shown in Figure 6. Spirulina also contains antioxidants, which are known to have significant effects on weight in obese diabetics persons, by increasing the body's need for energy while simultaneously preventing the production of adipocytes, and the enzyme lipase. Spirulina has been found to reduce lipid accumulation in the liver by inhibiting macrophage infiltration into visceral fat. Its low phenylalanine content is also reported to cause increases in the secretion of cholecystokinin, a hormone that suppresses appetite.

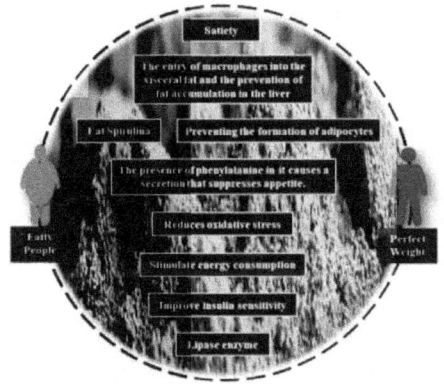

Fig. 6: Role of spirulina algae play in the process of weight loss (Source: AlFadhly et al., 2022)

While several other supplements have been evaluated and promoted for their lipid-lowering and weight-loss effects, the benefits derived from spirulina supplementation also extend to other health conditions, therefore making its use a cost-effective option. Clinical trials such as the one conducted by Yousefi and colleagues evaluated the effect of 2 g spirulina per day on 52 obese participants with body mass index (BMI) greater than 25 – 40 kg/m^2 who were restricted to a caloric diet for 12 weeks. They reported significantly lower body weight of -3.22±1.97 kg, a waist circumference of -3.37±2.65 kg, body fat of -2.28±1.74 kg, and BMI of -1.23±0.79 kg/m2. Additionally, the triglycerides values were reduced by -18mg/dL, while high-sensitivity C reactive protein levels were lower by -1.66±1.9ng/mL towards the end of the study period. Other studies have validated the ability of 1 – 2 g of spirulina per day for three months to reduce the body weight, BMI, and waist circumference of obese adults. A considerable rise in body weight, total protein, albumin, and hemoglobin levels has also been observed in diabetic rats given an aqueous extract of spirulina for a period of fifty days, indicating an improvement in general health conditions, and metabolic mechanisms brought about by efficient glycemic management.

It has particularly been observed that spirulina has more beneficial effects on weight and waist circumference when used for at least 12 weeks, while it has positive effects on BMI when used for a longer period. Contrarily, Huang, and coworkers reported an opposite effect when spirulina supplements were consumed at doses ranging from 1 to 19 g per day for a period ranging from 2 to 48 weeks. They reported an increase in overall body weight and BMI in non-obese persons. For example, compared to rats fed a control diet, rats given either pomegranate juice, spirulina extract, or both have been reported to record considerable rise in their body weights. The liquids stimulated hunger, which ultimately resulted in weight gain.

The Role of Spirulina in Reduction of Blood Cholesterol Level

Several studies have demonstrated the positive effects of spirulina on metabolic risk factors such as elevated blood lipids. Spirulina particularly lowers the serum levels of triglycerides and cholesterol linked to low-density lipoprotein (LDL), as well as having an indirect effect on total cholesterol and cholesterol linked to high-density lipoprotein (HDL). These properties have been attributed to its phycocyanin pigment, which as an antioxidant, protects cells and natural chemicals in the body from the damage caused by free radicals and oxidation. When consumed, spirulina also plays a part in the indirect modification of high cholesterol and total cholesterol levels. A rat study reported by McCarty and Kerna showed that oral phycocyanin boosted bile acid and cholesterol excretions—an effect that could be expected to lower LDL cholesterol levels, while several drugs used to lower LDL cholesterol, block cholesterol or bile acid reabsorption. It has been suggested that this effect is mediated by the antioxidant activity derived from intact phycocyanin. Functional foods that are abundant in spirulina can therefore be expected to aid LDL cholesterol control. Figure 10 provides a graphic of the roles that spirulina algae play in lowering cholesterol levels.

Fig. 10: The role of spirulina algae in reducing cholesterol levels (Source: AlFadhly et al., 2022)

Colla and colleagues monitored the total cholesterol, triglycerides, and HDL levels of hypercholesterolemic rabbits fed a diet that included 0.5 g of spirulina per day as a dietary supplement, and observed an association between the increase in the level of HDL in the blood and the decrease in the levels of total cholesterol, while there was no noticeable change in the levels of triglycerides. Spirulina can indeed limit the production of cholesterol, which is mediated the abundant gamma-linolenic acid in spirulina. Deficiencies of gamma-linolenic acid have been associated with the thickening of arterial walls, which could lead to high blood pressure. Specifically, the consumption of spirulina algae has been recommended for cardiovascular disorders such as high cholesterol and atherosclerosis, since it contains high levels of niacin which aids blood lipid balance.

The Role of Spirulina in Lowering the Risk of Diabetes

In addition to its impact on lipid metabolism, which is linked to diabetes, spirulina supplementation has also been used in the treatment of hyperglycemia during diabetes.

Diabetic conditions usually result in oxidative stress from reactive oxygen species, which may lead to the development of pancreatic-cell dysfunction. Several studies in rats and diabetic patients have confirmed that spirulina improves insulin sensitivity, and reduces blood glucose levels. This function has been credited to the presence of several bioactive components functioning to inhibit inflammatory activities, increase insulin sensitivity, inhibit gluconeogenesis, improve antioxidant activity, modulate gut microbiota composition, improve glucose homeostasis, and insulin receptor activation as shown in Figure 10. Spirulina-derived phycocyanin reportedly ameliorated diabetes in an animal study via activation of insulin signaling pathway and glucokinase expression in the pancreas and liver, in addition to modulating gluconeogenesis and apoptosis. Indeed, spirulina protein and amino acid constituents elevate blood glucose transport to the peripheral tissues and also stimulate insulin secretion from β-cell of the pancreas. Clinical trial studies have shown that spirulina supplementation at 2 g/day for 2 months could ameliorate fasting plasma glucose, postprandial glucose blood, and HbA1c levels in patients with type 2 diabetes.

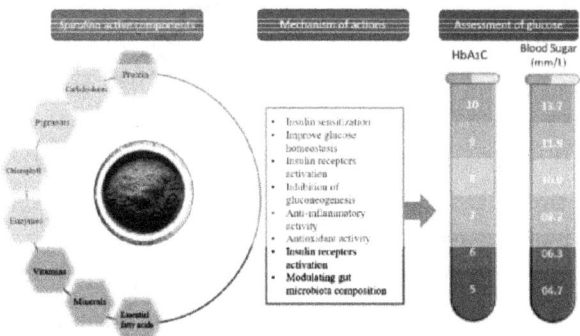

Fig. 10: Spirulina active components and anti-diabetic mechanism of action (Source: El-Sakhawy et al., 2023)

A recent study by researchers at the Shiraz University of Medical Sciences, Shiraz, Iran, however, reported insignificant effects of spirulina sauce on type-2 diabetic patients. They evaluated the effect of spirulina sauce on glycemic indices, lipid profile, oxidative stress markers, and anthropometric measurement in type 2 diabetic patients assigned 20 g/day spirulina sauce containing 2 g of spirulina for 2 months. They reported that among the lipid profiles, triglyceride, total cholesterol, and LDL were significantly decreased, while no significant change was recorded in HDL level. The hunger index was also significantly decreased, while fullness increased marginally. They did not however observe much change in total antioxidant capacity (TAC) body composition, and other anthropometric measurements, except waist circumference, which was reduced. They concluded that spirulina sauce may not be effective for glycemic control in type 2 diabetes, but it could be useful for controlling appetite, and ameliorating lipid profiles in diabetic patients. It is possible that the presentation of the spirulina as a sauce may have influenced the results. A larger sample size and longer follow-up duration may be needed to better elucidate the efficacy of such a spirulina sauce.

Anti-inflammatory Activities of Spirulina

One of the notable pharmaceutical applications of spirulina is its use in the treatment of inflammatory conditions, which is a contributing factor to numerous non-communicable diseases, including heart disease, diabetes, and arthritis. Studies have demonstrated that spirulina can modulate inflammatory responses in the body, potentially offering a natural alternative to conventional anti-inflammatory medications. The major pigment, phycocyanin found in spirulina as an antioxidant helps to fight oxidative stress by blocking the production of molecules that promote inflammation, thus conferring on spirulina an impressive antioxidant and anti-inflammatory property. The beneficial effects of *S platensis* in reducing obesity-associated chronic inflammatory state have also been associated with its intestinal activities through possibly its ability to regulate intestinal barrier function, or cause repairs in intestinal tissues damaged by high fiber diets (HFD).

A recent study at the Nanjing Medical University, Nanjing, China reported that *S. platensis* could largely reduce the relative amount of intestinal microbes such as *Proteobacteria* and the *Firmicutes/Bacteroidetes* ratio in fecal samples from HFD-fed rats, in addition to significantly reducing intestinal inflammation, as shown by decreased expression of myeloid differentiation factor 88 (MyD88), toll-like receptor 4 (TLR4), NF-κB (p65), and inflammatory cytokines. They also reported the ability of *S. platensis* to ameliorate the increased permeability, and decreased expression of tight junction proteins such as ZO-1, Occludin, and Claudin-1 in the intestinal mucosa. Therefore, *S. platensis* could alleviate chronic inflammation by modulating gut microbiota and intestinal permeability in rats fed a high-fat diet. Again, an inflammatory response associated with muscle damage during and after exercise has been associated with increases in inflammatory markers and intracellular proteins, and the release of cytokines like tumor necrosis factor to repair the damaged tissue. Such an inflammatory status usually causes muscle pain and functional decline, impairing the overall performance of athletes. Recent studies by Chaouachi and colleagues have shown that spirulina supplementation could prevent exercise-induced inflammation and skeletal muscle damage 24 hours after exercise in elite athletes.

Anti-Cancer Properties of Spirulina

S. platensis is a source of potent antioxidants such as spirulans, selenocompounds, phenolic compounds, phycobiliproteins, and chlorophyll. Since cell proliferation is strongly influenced by redox signaling, the antioxidant action of chlorophylls might account for the presumed anti-proliferative properties of *S. platensis*. It has been demonstrated that the high concentration of antioxidant molecules found in green alga is mostly responsible for their effectiveness as an alternative treatment for several non-communicable diseases such as cancer. Spirulina has therefore been shown to exert therapeutic effects against some types of cancer. Spirulina also boosts the stimulation of antibody and cytokine production, as well as natural killer cell activation. Additionally, it can augment the immune system, thereby playing a role in tumor growth inhibition. Spirulina exerts its effects on human myeloid progenitors and natural killer cells in either a direct or indirect manner, depending on the context. Again, the anti-cancer effect has been reported to induce mitochondrial dysfunction through the up-regulation of certain essential proteins that promote cell death similar to phosphorylation. Indeed, in cancer cells, an increase in the level of reactive oxygen species (ROS) usually results from metabolic activities, mitochondria malfunction, peroxisome activity, an increase of signaling through the mediation of the receptors, activation of oncogenes, and increased oxidase, cyclooxygenase, lipoxygenase, and thymidine phosphorylase activities. Figure 11 presents a schematic representation of the anticancer effects that may be linked to the consumption of spirulina algae.

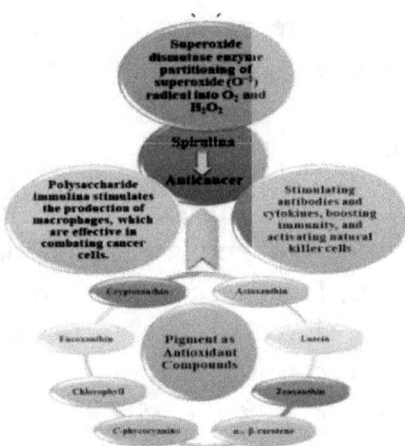

Fig. 11: Anticancer properties of the spirulina algae (Source: AlFadhly et al., 2022)

Existing studies have also documented the potency of several spirulina pigments and extracts in cancer treatments. For example, astaxanthin tincture has been reported to exhibit potent anti-inflammatory, and antioxidant capabilities, which can prevent or lessen the severity of several ailments, including cancer. Lutein, zeaxanthin, and beta-carotene are known to reduce the risk of developing premenopausal breast cancer, while cryptoxanthin and alpha-carotene have been shown to reduce the risk of developing cervical cancer. Again, chlorophyll tincture has been reported to have both antioxidant and anticarcinogenic activities, while fucoxanthin pigment has been proven to have antiobesity and anticarcinogenic effects. A recent study at Charles University, Prague, Czech Republic evaluated the anticancer effects of *S. platensis* and *S. platensis*-derived tetrapyrroles using an experimental model of pancreatic cancer. The anti-proliferative effects of *S. platensis* and its tetrapyrrolic components (phycocyanobilin (PCB) and chlorophyllin, a surrogate molecule for chlorophyll A, were tested on several human pancreatic cancer cell lines, and xenotransplanted nude mice. They reported a significant decrease in the proliferation of the human pancreatic cancer cell lines *in vitro* in a dose-dependent manner. The anti-proliferative effects of *S. platensis* were also shown *in vivo*, where inhibition of pancreatic cancer growth was evidenced since the third day of treatment. They therefore concluded that dietary supplementation with spirulina algae might enhance the systemic pool of tetrapyrroles, known to be higher in subjects with Gilbert syndrome characterized by defects in the processing of bilirubin by the liver.

Indian reports have also shown that administration of *S. fusiformis* at a dosage of 1 g/day for 12 months to tobacco chewers resulted in complete regression of oral leukoplakia lesions in 20 of 44 (45%) evaluated patients, as opposed to three of 43 (7%) in the placebo group. Discontinuation of further *S. fusiformis* administration for one year however resulted in nine of 20 (45%) complete responders developing recurrent lesions. Grawish and his team also reported the oral chemopreventive effects of the *S. platensis* extract, astaxanthin in experimental Syrian hamsters administered the extract at a dosage rate of 10 mg/day for 14 days. These studies indicate that both the spirulina algae and its extracts, especially astaxanthin carotenoid, have the potential to manage not only precancerous lesions, but also proper oral cancer. Other studies have also reported novel applications of *S. platensis* in managing an insidious and disabling diseases affecting the oral mucosa, such as Oral Submucous Fibrosis (OSMF). A chronic progressive pathology that alters the flexibility of the oral mucosa, ending in lockjaw, and a higher risk of developing squamous oral cancer, OSMF has been diagnosed as a common condition in India, southeastern Asia, and other parts of the world. A study by Hwang and colleagues was designed to assess the effectiveness of *S. platensis* together with different physiotherapeutic approaches in the treatment of OSMF. The *S. platensis* was administered at 500 mg twice daily for 3 months. All the subjects reported statistically significant amelioration in burning perception, mouth opening, tongue protrusion, and cheek flexibility, with no considerable side effects. The use of *S. platensis* for three months in treating the disease symptoms has also been found to elicit better clinical improvement in mouth opening, and reduction in burning perception than corticosteroid injections for the same period. These findings show that *S. platensis* could be used as an adjuvant therapy for the initial symptoms in subjects with OSMF.

The Role of Spirulina in the Treatment of Cardiovascular Disease

Several studies have demonstrated the beneficial effects of oral consumption of spirulina in preventing cardiovascular diseases through its ability to lower blood pressure and plasma lipid concentrations, especially triacylglycerols and low-density lipoprotein-cholesterol. It also indirectly modifies the total cholesterol and high-density lipoprotein cholesterol values. This decrease in total cholesterol and triacylglycerols, and increase in HDL levels is usually accompanied by a significant decrease in systolic and diastolic blood pressure. Contrarily, Haung and his team observed that supplementation with spirulina led to a substantial drop in diastolic blood pressure (DBP), but at the same time, a rise in systolic blood pressure, which is one of the indications of cardiovascular safety and metabolism in humans. These findings however suggest that spirulina has a beneficial influence on blood pressure through a mechanism of vesicular dilation. Specifically, an oral dose of 4.5 g per day of spirulina for six weeks has been associated with a drop in systolic and diastolic blood pressure. The high potassium and relatively low sodium content of the algae have been shown to play some role in the beneficial effects on blood pressure. Again, the C-phycocyanin pigment has been shown to stop blood platelets from initiating clotting by preventing Ca mobilization, and moderating the free radicals usually generated by platelets. The role of spirulina in preventing atherosclerosis has earlier been discussed.

Khan and colleagues in their series of studies have demonstrated that both the spirulina algae and its C-phycocyanin extract have preventive effects on drug-induced cardiac side effects as well as a protective effect during heart attacks in mice and rats.

Consumption of phycocyanin or spirulina algae enhanced the recovery of heart function in terms of decreased infarct size, attenuated lactate dehydrogenase and creatine kinase release, and suppression of ischemia-reperfusion-induced free radical generation. Sodium Spirulan, a novel polysaccharide isolated from spirulina by Japanese researchers, and a combination of spirulina phycocyanin and selenium have also shown cardioprotective qualities.

Effect of Spirulina on Aging

Aging has been defined as a progressive functional deterioration associated with frailty, disease, and death. It has also been defined as a progressive attenuation of the biological functions of the cell of an organism. It has been associated with stress and deterioration of immune responses, resulting in low-grade inflammation and increased susceptibility to infections, which collectively lead to severe disease conditions. Age-related liver changes such as an increase in the size and number of liver cells, and a reduction in mitochondrial number in the cells have been shown to significantly affect liver morphology, physiology, and oxidative capacity. The inflammation theory has indeed been applied to hepatic disorders since aging predisposes to hepatic functional and structural impairments, inflammation, and metabolic risk, which favor non-alcoholic fatty liver disease that can evolve into non-alcoholic steatohepatitis. Research evidence has also revealed extensive interplay between the gut microbiota, the immune system, and inflammatory pathways that influence aging, which are equally affected by genetic, environmental, and lifestyle factors. Recent studies have also shown that mitochondria play a crucial role in the aging process by producing reactive oxygen species (ROS). This role is supported by the free radical theory of aging, and the mitochondrial theory of aging. ROS accelerates cellular senescence by damaging various cellular components, such as DNA and proteins, and inhibiting the function of various organelles.

Specifically, the superoxide, produced in the mitochondria is the most reactive oxygen species rapidly converted by enzymes localized in the cytosol and mitochondria to hydrogen peroxide. Loss of the essential antioxidant enzymes localized in the mitochondria can also accelerate mitochondrial dysfunction and aging in cultured cells and animal models. Thus, these enzymes are critical to maintaining mitochondrial homeostasis and anti-aging. It has also been postulated that aging impairs the endoplasmic reticulum (ER) functions, which are critical in protein synthesis, indicating that maintenance of ER functions is also crucial for anti-aging strategies. Machihara and his team demonstrated in a recent study the ability of spirulina polysaccharide complex (SPC) to restore mitochondrial function, and collagen production by scavenging superoxide via the upregulation of the essential antioxidant enzymes in aging fibroblasts. They were able to link the process to inflammatory pathways, although the SPC did not upregulate the expression of most of the inflammatory cytokines produced as a result of the induction of lipopolysaccharide (LPS) in aging fibroblasts. Furthermore, the SPC can stimulate ER functions, indicating that the spirulina polysaccharide complex is a potential anti-aging substance that rejuvenates aging fibroblasts by increasing their antioxidant potential.

The aging process has also been shown to seriously affect the composition of the human gut microbiota, such that there is usually a decreased intestinal motility which results in a slower intestinal transit time that leads to constipation, reduction in bacterial excretion, and alterations in the fermentation processes in the gut. Counteracting these gerontological processes in the liver and other sites through gut microbiota modulation is a critical factor in healthy aging. Neyrinck and colleagues have shown that spirulina supplementation through the oral route caused changes in gut microbiota composition in old mice, with an increase in the relative abundance of the beneficial *Roseburia* and *Lactobacillus bacteria*. This suggests that the antimicrobial activity of spirulina targets other bacteria. The study also reported that spirulina supplementation reduced several hepatic inflammatory and oxidative stress markers that are usually upregulated in old mice. They concluded that oral administration of spirulina can modulate the gut microbiota, and activate the immune system, a mechanism that may be involved in enhanced hepatic inflammation in aged mice. Again, The abundant gamma-linolenic acid in spirulina can maintain healthy bones, alleviate back pain, prevent arthritis, osteoporosis, and kidney stone formation, and protect the teeth by ensuring that jaw bones remain strong in the elderly. Spirulina supplementation is also used in the enhancement of brain functions, alleviation of the symptoms of osteoporosis, enhancement of insulin metabolism, facilitation of the elimination of kidney stones, promotion of metabolic processes, defense against oxidative stress, and prevention of vitamin D deficiency in elderly people.

The Role of Spirulina in the Treatment of Neuro-degenerative Disorders (NDD)

Neuro-degenerative disorders have been linked to the deficiency of naturally occurring antioxidants and anti-inflammatory defensive mechanisms of the body. This deficiency predisposes nervous tissues and the brain to the damaging effects of free radicals, such as ROS and reactive nitrogen species (RNS), that play important roles in neurological disorders, such as Alzheimer's and Parkinson's diseases, multiple sclerosis, inflammatory lesions, and aging. Microglial cell activation has particularly been associated with the pathogenesis of neuro-degenerative diseases, including Parkinson's disease. The activated microglia has however been reported to exhibit both neuroprotective and neurotoxic effects, although persistent activation of microglia eventually leads to neuronal death due to neurotoxicity. The neuroprotein, alpha-synuclein has also been implicated in the etiology of Parkinson's disease through its involvement in many neuro-transmission processes. Extracellular release of alpha-synuclein has been associated with increases in pro-inflammatory cytokine and ROS production, which damages the afflicted neurons and surrounding nervous tissues.

Natural products, such as the spirulina blue-green algae, are believed to help reverse these effects through their anti-inflammatory/anti-oxidant properties. Some studies have specifically shown that diets enriched with spirulina could enhance the recovery of dopamine-positive fibers, and neuron lesions in Parkinson's disease models by preventing the upregulation of certain major histocompatibility complex molecules (MHCII) expression one month after the lesion. Piovan and colleagues reported that spirulina extract that blocks lipopolysaccharides, helps to control the activation of microglia and prevent the occurrence of neuro-inflammation. The neuroprotective role of spirulina in mitigating the effects of spinal cord injury, the spinal cortical tracts, and behavioral recovery in injured laboratory rats given 180 mg/kg spirulina has also been reported. Pabon and her team have also observed a decrease in activated microglia in the rats that received a spirulina-enhanced diet, indicating the potential use of spirulina to promote neuroprotection in animal models of Parkinson's disease. Again, polycaprolactone spirulina nanofiber mat used in tissue engineering has proven to be effective against CNS injury, as it decreases the astrocyte activation after ischemic stroke which in turn, could reduce inflammation induced by the abnormal increase in the number of astrocytes. Animal studies have also shown that daily consumption of spirulina suppresses photostress-induced retinal damage, and prevents vision loss in mice. The abundant gamma-linolenic acid in spirulina can maintain healthy bones, alleviate back pain, prevent arthritis and osteoporosis, avoid kidney stones, and protect teeth by ensuring that jaw bones remain strong.

Conclusion

Spirulina supplementation has several benefits such as treatment of cardiovascular diseases, and neuro-degenerative disorders, enhancement of the immune system, reduction of blood cholesterol, and anti-inflammatory anti-cancer, anti-diabetic and anti-aging functions. Other function include the enhancement of brain function, alleviation of the symptoms of osteoporosis, enhancement of insulin metabolism, facilitation of the elimination of kidney stones, promotion of metabolic processes, defense against oxidative stress, and prevention of vitamin D deficiency.

Prof Ifeanyi Charles Okoli

USES OF SPIRULINA IN ANIMAL PRODUCTION

(Image source: Wiseman Spirulina)

Introduction

Economic and environmental issues are becoming increasingly important to the search for alternatives that will help reduce the competition between humans, intensively farmed livestock and poultry for plant-based foods such as legumes, grains, oilseed cakes, and industrial by-products. Microalgae with their richness in biologically active substances, and limited usage as human food are promising nutrient sources for animal feeding. For example, spirulina cultivation is eco-friendly, since it does not require the use of fertile lands, and is also characterized by rapid growth and turnover. The energy input and water needed to produce a kilogram of spirulina are much less than soya and corn proteins. This cost-effective production and high nutritional value of spirulina make it an ideal protein-rich animal feedstuff for improving the performance, feed conversion, and product quality of several livestock and poultry species.

Thus, many studies have highlighted the value of spirulina in the improvement of growth performance, health, milk production, egg production and meat quality, and reduction in methane emissions from farmed animals. Some studies have also reported that supplementing livestock feed with spirulina can improve meat quality characteristics such as increased tenderness, juiciness, and higher omega-3 fatty acids, and other beneficial nutrients. Another important benefit of spirulina supplementation of animal feed is its ability to reduce the emission of the greenhouse gas, methane from livestock units. Reports have shown that when added to the cattle diet, spirulina improves feed digestion while reducing the amount of methane produced during the digestive process by up to 40 percent.
 Similarly, a study by Roohani and colleagues has shown that spirulina could be a strategic option for sparing expensive fish meal in aquaculture feeding, since its supplementation at 6 and 8 percent levels resulted in better growth performance and fish fillet quality than the control.

Prof Ifeanyi Charles Okoli

The versatile applications of spirulina makes the algae a key resource for sustainable agriculture and aquaculture. As the agro-industry seeks to innovate and implement more eco-friendly practices, spirulina inclusion into animal feed formulations should be seen as a viable solution that aligns with the principles of circular economy and sustainable development. By harnessing the potential of spirulina, stakeholders across various agricultural sectors can contribute to a more resilient food system, ultimately supporting global efforts to enhance food security and mitigate climate change.

SPIRULINA AS ANIMAL FEED SUPPLEMENT

Rich in proteins, vitamins, and essential fatty acids, spirulina is increasingly being incorporated into animal feed formulations to enhance growth rates, improve feed efficiency, and bolster the immune systems of livestock and aquaculture species. Its high digestibility and nutrient density make it an ideal candidate for fortifying diets, particularly in intensive farming systems where conventional feed ingredients such as cereals, tubers, and legumes may fall short of providing the required nutrients. Thus, incorporating spirulina into such feeds has significant advantages in various livestock sectors, including poultry, swine, ruminants, and aquaculture. Available studies have reported significant improvements in weight gain, better feed conversion ratios, and enhanced overall health in animals fed diets supplemented with spirulina. Furthermore, the rich antioxidant content of the algae has been shown to mitigate oxidative stress commonly experienced by animals under intensive production systems. These benefits not only support animal welfare but also contribute to more sustainable farming practices by reducing reliance on antibiotics, and other harmful growth promoters.

Effects of Dietary Spirulina Supplementation on Poultry Performance

Spirulina is increasingly being used as a feed additive in poultry production based on research, and field reports that it can improve the growth performance and health of poultry. The high protein content (up to 70 percent) is being exploited for growth promotion. The high vitamin, mineral, and antioxidant contents have been shown to improve the overall health and resistance of birds to endemic diseases. Some studies have reported that spirulina supplementation at 0.5 or 1 percent in broiler chicken diets helped them to overcome the effects of heat stress on their production performance, while also improving their carcass dressing, breast and leg percentages. Park and coworkers also reported that due to its high content of antioxidant compounds, spirulina inclusion in broiler chicken diets enhanced the growth efficiency, and nutrient digestibility, increased antioxidant and enzyme activities, modulated cecal microflora, and reduced the emission of noxious gases from the excreta. The effect of spirulina on the growth and health of chickens is shown in Table 6.

Commercial spirulina poultry feed supplement (Image source: IndiaMART)

Table 6: The effects of spirulina on the growth performance and health of chickens

Parameter	Summary of results
Growth	Growth rates declined in 3-week-old chicks fed Spirulina levels of 10% and 20% of diet
	Body weights of chicks fed Spirulina levels of 11.1% and 16.6% of diet were not different from the control group, receiving groundnut cake
	Broilers fed Spirulina levels of 140 and 170 g/kg of diet and vitamin and mineral premixes omitted had no difference in dressing percentage compared to those receiving fishmeal or groundnut cake
	Broilers fed Spirulina levels of 0, 40 or 80 g/kg of diet for 16 days did not significantly differ in body weights
	Broilers fed Spirulina levels of 40 g/kg of diet had greater muscle redness and yellowness than the control group
	White Leghorn and broilers fed Spirulina levels of 0, 0.001, 0.1, 1 and 10 g/kg of diet had comparable body weights after 7 weeks
Health	Chicks fed Spirulina levels of 10 g/kg of diet had increased NK cell activity compared to the control group, showing an enhanced disease resistance potential
	Chicken phagocytic activity had an incremental linear increase with increasing dietary Spirulina levels of 0.5%, 1% and 2% of diet
Product quality	White Leghorn hens' egg total cholesterol levels were reduced when diets contained 150 g flaxseeds + 200 mg vitamin E + 3 g Spirulina per kg diet
	White Leghorn layers, aged 32 weeks, fed 20% whole flaxseeds and 5% Spirulina (w/w) produced eggs with higher levels of linoleic acid with less cholesterol
	Egg yolk colour score was higher in layers fed flaxseed diets with 5% Spirulina (w/w) compared to those on a flaxseed diet 0% w/w
	Optimal egg yolk pigmentation was obtained by feeding Spirulina levels of 1% of diet, when diet is otherwise free of xanthophylls
	Egg yolk carotenoids pigment and omega 3 fatty acid levels increase when White Leghorn hens fed 150 g flaxseeds + 200 mg vitamin E + 3 g Spirulina per kg diet

(Source: Holman and Malau-Aduli, 2013)

Another studies by Evans and colleagues showed that dried full-fat spirulina algae had an energy value equal to 90 percent of the energy in corn, and 76 percent crude protein. They also reported that up to 16 percent of the dried algae could be incorporated into a broiler starter diet without negative effects on the performance of chicks. Other studies have also reported optimal performance in broilers fed a diet containing 12 percent dehydrated spirulina as replacement for the other protein sources. An additional benefit derived from spirulina supplementation is the increase in the yellowness of the muscles, skin, fat, and liver components of broiler meat, thus making the meat more desirable to customers in many countries. Incorporation of 1 to 2 percent dried spirulina powder in laying hen diets has also been reported to cause a significant increase in egg yellow color values compared to the control group, although there were no significant changes in laying performance.

Effects of Dietary Spirulina Supplementation on Pig Performance

Several studies have shown that spirulina could be a good protein source in pig diets. Indeed, Grinstead and coworkers in their early study reported that pigs fed 20 g spirulina per kg feed recorded no significant improvements in their feed intake and daily gains.

Pig growth responses to dietary spirulina supplementation have however been inconsistent as depicted in Table 7. Some studies have however reported up to 9 percent increase in growth rates among weanling pigs fed diets supplemented with spirulina. A more recent study at the Universidade de Lisboa, Lisboa, Portugal reported impairments in the growth performance of post-weaning piglets fed diets containing 10 percent spirulina, which is mediated by an increase in digesta viscosity and lower protein digestibility, as a consequence of the resistance of microalga proteins to the action of endogenous peptidases. Contrarily, Neumann and coworkers observed that soybean meal could be replaced entirely by spirulina (with appropriate lysine supplementation) in swine diets without compromising overall protein quality.

They equally observed that supplementing a high quantity of histidine with lysine, methionine, and threonine improves the protein quality of swine diets containing spirulina. Based on these reports there is the need for appropriate amino acid supplementation of pig diets containing high levels of spirulina. Studies on the effects of spirulina supplementation on the sensory characteristics of pig meat have also reported increases in the overall odor of loin meat, and astringent aftertaste in pork loin, although the changes were marginal and should therefore not be a drawback to product quality.

Table 7: The effects of spirulina on the growth performance and health of pigs

Parameter	Summary of results
Growth	Crossbred weanling pigs fed Spirulina levels of 1.5% and 3% of diet had higher growth rates compared to the control group
	Weanling pigs fed Spirulina pelleted diets had decreased average daily gain (ADG), while those receiving Spirulina in meal diets had improved ADG
	ADG in pigs fed Spirulina levels of 2% of diet was greater than in the control group, during days 14–28 post-weaning
	Pigs fed Spirulina levels of 14% of diet had similar growth as those fed skim milk powder
	Increasing Spirulina levels in pig diets (0.5%, 1% and 2% diet) showed only a numerical increase in ADG
Fertility	Boars fed BioR (extracted from Spirulina) at 1.5 ml/day had increased ejaculate volume and spermatozoa mobility compared to a control group

(Source: Holman and Malau-Aduli, 2013)

Effects of Dietary Spirulina Supplementation on Ruminant Performance

Ruminant capacity to process natural microalgal material makes them ideal for dietary spirulina supplementation. For example, approximately 20 percent of dietary spirulina bypasses degradation in the fore stomach, and reaches the abomasum where it undergoes direct absorption, indicating a potential positive effect of adding the algae to the rumination feed. Table 8 highlights the effects of spirulina supplementation on the growth performance and health of ruminants. An Egyptian study evaluated the impact of 1 and 2 g dry *S. platensis* supplementation per day in the rations of growing Friesian calves on feed intake, rumen fermentation activity, digestibility, growth performance, and economic efficiency. The calves fed the spirulina supplementee diets recorded better nutrient digestibility, feed intake, rumen fermentation activity, blood biochemical parameters, body weight gain, and feed conversion ratio than the control calves. While the feed cost per kg live weight gain was significantly lower for the supplemented spirulina rations, the total body weight gain, net revenue, and economic efficiency were significantly higher for calves fed spirulina rations than those of the control ration. Spirulina supplementation at 1 g/10 kg b.wt./day in the diets of fattening lambs for 35 days was also reported to improve daily weight gain, final live body weight, feed intake, and feed conversion rate compared to the control.

Table 8: The effects of spirulina on the growth performance and health of ruminants

Species	Parameter	Summary of results
Cattle	Growth	Dairy cows fed 200 g Spirulina daily were 8.5–11% fatter than the control group, evaluated using body condition score
	Productivity	Dairy cows fed 200 g Spirulina daily produced more milk than the control group
		Cows fed Spirulina levels of 2 g/day (w/w) produced more milk than the control group
		Spirulina levels of 0.15% of diet resulted in decreased rumen degradability of dietary crude protein
	Product quality	Milk from cows fed Spirulina levels of 2 g/day had greater average milk fat, protein and lactose than controls
		Milk saturated fatty acid levels decreased, while mono- and polyunsaturated fatty acids increased when crossbred Holsteins were fed Spirulina at 40 g/day
		Spirulina fed at 2 g/day to dairy cows reduces the somatic cell counts
Sheep	Growth	6-month-old lambs fed Spirulina levels of 10% (w/w) had greater liveweights than those given 20% (w/w) and the control group
		Lambs body condition scores incrementally higher in lambs fed Spirulina levels of 10% and 20% (w/w) compared to controls
		Lambs fed cow milk enriched with 10 g/day Spirulina had higher liveweights and growth rates during 15–30 days old than the control group
		Pregnant ewes fed pellets containing 2 g Spirulina ad libitum produced newborn lambs with higher weights and average daily gains than those from control treatment ewes

(Source: Holman and Malau-Aduli, 2013)

In addition, the hemoglobin concentration, total leukocytic count, serum globulin, vitamin A, and glutathione (GSH) concentrations were increased, while the serum enzymes, cholesterol, and glucose concentrations decreased below the control values. Other studies that compared the effects of low and high dietary spirulina supplementation levels on the reproductive performance of Australian crossbred sheep reported increased lamb production, associated with higher levels of long-chain polyunsaturated fatty acids at low supplementation levels. At high levels, spirulina supplementation decreased the intramuscular fat by decreasing the fat deposition in subcutaneous adipose tissue.

Spirulina also enhances wool production and quality. Thus, it can be supplemented in the diet to enhance the profitability of wool-producing animals. One of the primary effects of spirulina on wool growth is the enhancement of keratin production. Keratin is the fundamental protein in wool fibers, and its synthesis is influenced by the availability of amino acids in the diet. Spirulina is particularly high in lysine and methionine, two essential amino acids for keratin production. Spirulina also enhances the strength and durability of the wool fibers, thereby making them more desirable products that can fetch higher market prices. The inclusion of spirulina in sheep diets can also positively influence reproductive performance by enhancing fertility rates, and improving the quality of sperm and eggs. Incorporating spirulina into goat diets can also lead to improved feed efficiency. This is because goats that consume spirulina tend to have better nutrient absorption due to its rich nutrient profile. This means that less feed may be required to achieve the same or better production outcomes, thus lowering overall feed costs for farmers. Improved feed efficiency also translates into better weight gain, and milk production, thus making spirulina a cost-effective supplement for goat farming.

Effects of Dietary Spirulina Supplementation on Rabbit Performance

Several reports also exist on spirulina supplementation of rabbit diets. Generally, dietary supplementation of spirulina has been shown to influence the feed consumption, growth, and carcass yield of rabbits. Even at levels as low as 1 percent total dry matter, spirulina improved the crude protein digestion of rabbits fed both low and high-fat diets, indicating that the algae may be useful in high-fat and energy-basal rabbit diets. For example, Dalle Zotte and colleagues showed that dietary inclusion of 3 percent spirulina in growing dwarf rabbit diets for 14 weeks did not alter their growth performance, and energy or nutrient digestibility. Spirulina supplementation in the diet of growing rabbits has similarly been reported to have produced no change in their body weight, weight gain, feed consumption, mortality, and morbidity, with the only significant effect being on the feed conversion ratio which decreased below the control value. Rabbit meat quality has also been shown to improve when rabbits receive dietary spirulina. For example, some studies have identified dietary spirulina as a causal factor for increasing GLA and PUFA ratios within rabbit muscle lipid contents, which are indications of improvements in the oxidative stability of the rabbit meat and consumer preference. Rabbit health has also been found to improve with dietary spirulina since rabbits receiving the algae in their diets had high oxyhemoglobin levels. Table 9 highlights the salient effects of spirulina supplementation on the growth and health of rabbits.

Table 9: The effects of spirulina on growth performance and health of rabbits

Parameter	Summary of results
Growth	Final weight and weight gain did not differ between rabbits fed *Spirulina* levels of 0%, 5%, 10% or 15% of diet
	Feed intake of rabbits fed *Spirulina* levels of 5% and 10% of diet was greater than the control and 15% groups
	Rabbits receiving *Spirulina* levels of 1% of diet had increased crude protein digestibility in both low- and high-fat diets
	Spirulina levels of 10% of diet resulted in high feed intake compared to control group
Health	New Zealand White rabbits fed a high-fat diet and supplemented *Spirulina* levels of 10 g/kg of diet had reduced reactive oxygen species and oxidative stress
Product quality	Γ-Linoleic acid content in the perirenal fat and meat tissue in rabbits increased with *Spirulina* levels of 5%, 10% and 15% of diet

(Source: Holman and Malau-Aduli, 2013)

Effects of Spirulina Supplementation of Aquaculture Feed

In aquaculture, spirulina has gained recognition as a vital feed supplement for various species of fish and shrimp. Its utilization is particularly advantageous for the fry and fingerling stages, where nutritional quality is crucial for survival and growth. Incorporation of spirulina into aquafeeds can improve pigmentation, enhance disease resistance, and positively influence the overall health of aquatic species. As the global demand for seafood continues to rise, incorporating spirulina into aquaculture diets represents a sustainable approach to meeting nutritional needs, while minimizing environmental impacts associated with traditional fish feed sources. The benefits therefore, extend beyond improved animal health and productivity since the cultivation of spirulina requires significantly less water, and land compared to traditional feed crops, making it a more sustainable alternative in the face of climate change and resource scarcity. Additionally, Spirulina farming can play a role in bioremediation, as it can absorb excess nutrients and pollutants from wastewater, thus contributing to cleaner water systems, while simultaneously serving as a valuable feed resource.

Spirulina for aquafeed formulation (Image source: Proalgaetech)

A study on the complete substitution of fish meal with spirulina in an isocaloric and amino acid-balanced diet, for rainbow trout reported no significant effects on their growth parameters and feed conversion. In two feeding trials conducted by Rosenau and colleagues with rainbow trout, brook trout, brown trout, and African catfish, on the acceptance, and performance of an experimental diet containing spirulina, the overall, acceptance of the diets was found to be high across all species except brown trout. The authors hypothesized that this species may have rejected the spirulina diet due to an aversion to unfamiliar flavors. Again, the overall digestibility of the diet was high, while the feed conversion ratio was increased for spirulina-fed rainbow trout and brook trout resulting in significantly lower growth rates in all species.

Use of Spirulina in Pet Care

Spirulina has also been considered an algae of veterinary importance, because of its benefits as a super food, and functional food for companion animals. Although spirulina is frequently promoted as an immunity and skin booster, the amount of spirulina in commercial pet foods, that will enhance the immunity and visible health outcomes in dogs and cats are not properly documented. Satyaraj and colleagues in their recent study however reported that dogs fed diets supplemented with spirulina demonstrated enhanced immune status in terms of higher vaccine response and levels of fecal immunoglobin A, as well as a significant increase in gut microbiota stability. Numerous research reports exist on the use of spirulina in the management of inflammation of the gut, liver, joint, and kidney diseases and a variety of other conditions in horses. Spirulina is probably best known for its ability to boost the immune system and has therefore been routinely used to manage obstructive pulmonary disease/asthma, seasonal respiratory allergies, skin allergies, and poor immune function in horses.

Spirulina for pet care (Image source: Versatile sea algae)

Conclusion

The value of spirulina in the improvement of growth performance, health, milk production, egg production and meat quality, and reduction in methane emissions from farmed animals has been highlighted. Feeding trials conducted on the use of spirulina in animal production have yielded valuable insights that can significantly influence its application in farming practices. These trials have demonstrated the potential of spirulina as a nutritional supplement that enhances growth rates, improves feed efficiency, and boosts overall health in livestock. Trials indicate that a moderate inclusion rate, typically 5 to 10 percent of the total feed, maximizes growth without negatively affecting feed palatability. Farmers have also reported noticeable improvements in the weight gain of their animals, which translates to better market value and profitability. The data collected from these trials underscore the importance of incorporating spirulina into animal diets to optimize their production potential.

USES OF SPIRULINA IN CROP PRODUCTION

Harvesting of spirulina (Image source: FFredom.com)

Prof Ifeanyi Charles Okoli

Introduction

The incorporation of spirulina in agricultural practices has shown significant benefits for both crop yield and quality, positioning it as a valuable resource for farmers and agricultural researchers alike. Spirulina algae is rich in nutrients, including proteins, vitamins, and minerals, which can enhance soil fertility and improve plant health. When used as a bio-fertilizer or soil amendment, spirulina contributes to a more balanced nutrient profile in the soil, promoting robust crop growth and increasing overall productivity. This natural approach not only supports higher yields but also fosters sustainable farming practices that can reduce dependency on synthetic fertilizers. In addition to enhancing soil quality, spirulina can directly influence the nutritional value of crops. Studies have demonstrated that crops treated with spirulina exhibit improved nutrient density, including higher concentrations of essential amino acids, vitamins, and antioxidants. This is particularly important for addressing food security, and nutrition in a world increasingly challenged by malnutrition and dietary deficiencies. By integrating spirulina into crop management strategies, farmers can produce food that is not only more abundant but also more beneficial for consumers, thus, contributing to better public health outcomes. The use of spirulina extends beyond individual crop benefits to impacting the entire agricultural ecosystems. Its application could improve resilience against pests and diseases, and also reduce the need for chemical pesticides. Spirulina contains bioactive compounds that can enhance plants' natural defense mechanisms, allowing them to better withstand environmental stresses. This bio-control aspect not only protects crop yield, but also aligns with the growing demand for organic and environmentally friendly farming practices. As consumers increasingly prioritize sustainably sourced products, adoption of spirulina in agricultural practices could enhance marketability, and consumer trust in agricultural products.

Several studies have shown that the application and processing methods of spirulina biomass in agriculture are relatively similar, and include direct inoculation with living cells and different treatments such as mechanical/physical extraction (e.g., autoclaving, drying, grinding, lyophilization, heating with water, sonication, (supercritical CO_2), chemical extraction (alkalis or acids), and enzymatic extraction (e.g., with proteases). The high content of free amino acids, as well as macro and micro-elements in spirulina which are absorbed relatively faster than soil nutrients, in addition to the presence of some growth-promoting substances directly absorbed by plant leaves, explains the use of spirulina as a bio-stimulator in organic farming systems. There is however limited research information and farmers' awareness about the benefits of spirulina as a better bio-fertilizer than chemical fertilizer for the production of different crops. Layam and colleagues have also reported that micro-nutrients such as iron and zinc in green gram (*Amaranthus gangeticus*) plants were increased by soaking the seeds in a solution of 5 g of spirulina in 100 ml of sterile water, or by planting them in a mixture of organic manure and spirulina. Zinc deficiency, a major global health problem could therefore be addressed effectively through such a spirulina bio-fortification approach. Additionally, spirulina can produce metabolites such as phytohormones, and polysaccharides, that contribute positively to agricultural production. The algae has also been used as a bio-control agent or bio-pesticide, due to its ability to effectively control or suppress the growth of pathogens such as fungi, bacteria, and nematodes. This is achieved through the activities of biocidal compounds like benzoic and majusculonic acids, and hydrolytic enzymes present in the algae. The knowledge about the potential value of spirulina in crop production is however limited or even lacking in many developing countries, and sub-Saharan Africa, even though it could be easily cultivated in most of these countries. Figure 11 highlights the major uses of spirulina microalgae in crop production, while Table 10 shows the major bioactive compounds identified in spirulina and their roles in agriculture.

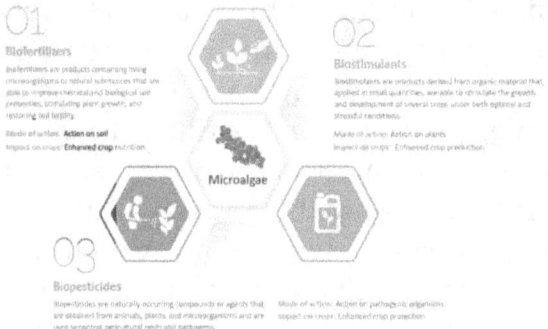

Fig. 11: The activities spirulina in crop production (Source: Gonçalves et al., 2021).

Table 10: Bioactive compounds identified in spirulina and their roles in agriculture

Bioactive Compounds	Biological Activity	Role in Agriculture
Phenolic compounds	Antioxidant, antibacterial and antifungal	Crop protection against pathogens and various biotic and abiotic stress conditions
Carotenoids	Antioxidant, anti-inflammatory and anticancer	Crop fortification; Soil bioremediation and fertilization; Crop protection against biotic and abiotic stress conditions
Terpenoids	Antioxidant, antibacterial and anticarcinogenic	Crop protection against insects, bacteria, and other organisms; Attraction of pollinators; Stimulation of plant growth and development
Polysaccharides	Antioxidant, anti-inflammatory, antibacterial; anticoagulant and anticancer	Crop protection against biotic and abiotic stress conditions; Improvement of soil quality; Stimulation of plant growth
Free fatty acids	Antioxidant, antifungal, antiviral, antibiotic, and anticarcinogenic	Crop protection against various biotic and abiotic stress conditions
Phytohormones	Chemical messengers	Crop response to stress conditions; Regulation of cellular activities in crops; Stimulation of plant growth

(Adapted from Gonçalves et al., 2021)

Spirulina as a Natural Bio-Fertilizer

Since the mid-20th century, soil application of dry biomass (bio-inoculation) of different cyanobacteria, initially called "legalization" was shown to improve the growth, health, and yields of various crops. Microalgae such as spirulina are increasingly being used as bio-fertilizers in many countries due to their rich content of micro and macro-nutrients, phytohormones, and bioactive compounds, that confer positive biochemical effects on the soil ecosystem, through the interactions between the crops and the soil microbiome. Microalgae also possess remarkable capabilities for nutrient recovery through their biochemical processes that enable them to grow, and efficiently assimilate phosphorus and nitrogen, even in nutrient-limited environments. On application, they can enrich the soil with these nutrients. In the soil, spirulina releases its biomass and organic compounds, thereby improving the soil structure, moisture-holding capacity, and overall fertility. For example, in India, blue-green algae are grown in shallow earthen ponds and when the water evaporates, the dried algae are scooped up and sold to rice farmers as a natural nitrogen source, which is only one-third the cost of chemical fertilizer and increases annual rice yield by an average of 22 percent. When used as the sole nitrogen source, spirulina algae application has been shown to give the same benefit as 25 to 30 kg of chemical nitrogen fertilizer per acre. Its use also allows the reductions in the rate of inorganic fertilizer application. Figure 12 highlights the bio-fertilizer potential of microalgae.

Fig. 12: Bio-fertilizing potential of microalgae on maize (Source: Dineshkumar et al., 2019)

The application of spirulina as bio-fertilizer has particularly been shown to stimulate plant growth and development, especially in terms of improvements in germination rates and plant characteristics, such as increase in root length, number of leaves, and leaf area, among others. Early Indian studies reported that the application of combinations of algae and other fertilizers such as the N_2-fixing cyanobacterium, *Aulosira fertilissima*, *S. platensis*, and the chemical fertilizer, diammonium phosphate (DAP) in potted tomato plants at the rate of four times at seven days interval, resulted in improved agronomic characteristics and yield. Specifically, the highest plant fresh weight (290 g/plant), number of leaves (127/plant), number of flowers (29/plant), number of fruits (37/plant) and fresh weight of fruits (71 g/plant) were achieved with the application of a combination of 2.25 g *Aulosira*, 2.25 g spirulina, and 0.50 g DAP in each pot, which represented a 522 percent increase in number of fruits, and a 977 percent increase in yield over the control.

It has been established that biological nitrogen sources are usually more beneficial to crops than inorganic nitrogen, since they also release carbon components and other nutrients, which enhance plant growth. This has recently become an important factor in sustainable agriculture because the depletion of soil organic carbon is a significant form of degradation in croplands, leading to decreased soil quality and fertility. Thus, spirulina as a fertilizer has the added advantage of containing both nitrogen and other macro- and micro-nutrients which are slowly released under normal conditions to increase soil fertility. It can assimilate organic carbon into its biomass through photosynthesis, and can also release exopolysaccharides (EPS), which help as a carbon source and carbon sequestrant, thereby enhancing soil aggregation and stabilization. The EPS also serves as a reservoir for water storage during water scarcity, and remains metabolically active when hydrated. Cyanobacteria such as spirulina are also increasingly being used as sustainably efficient tools in the restoration of degraded soils affected by excess salts.

Again, spirulina influences soil microbial populations in terms of diversity, community composition, and activity, through its intrinsic capacity to produce a wide range of bioactive metabolites such as phenolic compounds, carotenoids, terpenoids, polysaccharides, free fatty acids, and phytohormones. These bioactive substances have also a positive impact on plant growth and control of pests and pathogens. Spirulina has equally demonstrated a significant capacity to remove wastewater pollutants such as nitrates and phosphates by as much as 93 and 86 percent respectively, thereby making them excellent candidates for soil bioremediation. The algae biomass also contains other nutrients such as potassium, magnesium, sulfur, and iron which are essential micro-elements for plant growth and development, through their involvement in redox reactions that play significant roles in plant metabolism. Spirulina can also promote the solubilization of other important nutrients, such as phosphorus, zinc, copper, and iron, thus making its use an effective strategy for improving soil nutrient availability to plants.

Spirulina extracts or suspensions have also been used as foliar fertilizers to supply the plants with the necessary nutrients by directly spraying the aqueous spirulina solution onto the leaves, which absorb these nutrients through their cuticles and stomata. Youssef and colleagues reported that *S. platensis* applications through foliar spraying, and soil drenching methods represent a promising approach to enhancing the growth and productivity of chia plants (*Salvia hispanica* L.) under alkaline stress conditions. Dalia and Sabree Kh, sprayed pea plants with either a 10 or 15 percent algae extract and reported improvements in vegetative development compared to the control. Other studies have also shown that fenugreek plants sprayed with an extract of *S. platensis* at two different concentrations (2.5 and 5.0 g L-1), on analysis recorded higher N, P, and K contents than the untreated plants. The application of spirulina in the soil and by foliar spraying of the extract, therefore, enhances the nutrient uptake by crops. Gharib and Ahmed investigated the response of rosemary (*Rosmarinus officinalis* L.) plants to foliar application of *S. platensis* at 0.0, 0.1, 0.2, and 0.4 percent, soil irrigation with heavy metals (Cd nitrate, Pb acetate, and Cd + Pb, each at 100 ppm), and *S. platensis* at 0.1 percent + heavy metals. They reported that *S. platensis* significantly improved the growth parameters, oil yield/fed, photosynthetic pigments, and activities of the enzymes, superoxide dismutase (SOD), glutathione reductase (GR), catalase (CAT), and polyphenol oxidase (PPO), with a maximum promoting effect at 0.2 percent algal extract. Specifically, *S. platensis* at 0.1 percent foliar application significantly increased growth parameters, oil content, photosynthetic pigments, and the activity of non-enzymatic and enzymatic antioxidants, while, slightly reducing the translocation factor (TF) of Cd and Pb, alleviated the membrane lipid peroxidation, and significantly lowered the content of malondialdehyde, hydrogen peroxide, and indole acetic acid oxidase (IAAO) activity in the heavy metal (Cd, Pb, and Cd+Pb)-treated rosemary plants. They concluded that foliar spray with *S. platensis* may be a viable novel strategy for agricultural improvement of medicinal and aromatic plants under normal and heavy metal (Cd, Pb, or Cd+Pb) stress.

The use of spirulina as a bio-fertilizer has however, some limitations, one of which is the high cost associated with its production, and the extraction of the metabolites of interest. Although cultivation is widely recognized as the primary cost contributor for algal-based products, harvesting and dewatering microalgae biomass are equally significant factors impacting the total costs. Again, microalgae such as spirulina can accumulate heavy metals and other contaminants when cultured in wastewater, indicating that their use as bio-fertilizers could potentially introduce these contaminants into the agricultural soil, thereby posing risks to crop growth and human health. Variations in the nutrient composition of the algae bio-fertilizer is another limitation that poses a challenge to maintaining consistent nutrients and is usually influenced by the characteristics of the wastewater used for the algae cultivation.

Spirulina as a Plant Bio-Stimulant

Plant bio-stimulants have been defined as fertilizers that function by stimulating plant nutrition processes independent of the product's nutrient content, with the sole aim of improving one or more of the characteristics of the plant, or the plant rhizosphere. These characteristics include nutrient use efficiency, tolerance to abiotic stress, quality traits, and availability of confined nutrients in the soil or rhizosphere. Plant bio-stimulation is usually achieved through the actions of bioactive compounds which are effective on plants at significantly lower concentrations compared to macro-nutrients. For example, the application of an *S. platensis* extract alone or in combination with the nitrogen-fixing bacterium *Pseudomonas stutzeri* in the presence of different doses of nitrogen fertilizer, has been used to enhance the growth, and productivity of onions under field conditions. In addition to their effects on plant growth, some bio-stimulants, especially algae extracts have been shown to trigger biochemical processes that result in the accumulation of important metabolites, with consequent improvement of qualitative traits of the final products. For example, foliar applications of different protein hydrolysate, seaweed, and plant extracts, were found to improve the commercial features and nutritional qualities of tomatoes and grapes. Foliar application of bio-stimulants has also been used to reduce the level of undesirable components such as nitrates in greenhouse-grown vegetables. Foliar applications of *S. platensis* extracts have an effect in modulating the nutritional and functional properties of the final marketable product. For example, application of two different concentrations of spirulina algae extract has been used to markedly increase both the oil and NPK content of fenugreek seeds, with the highest increase (+90 percent) in oil content recorded in the seeds treated with 5 g L^{-1} of spirulina extract. A similar increase in oil (+77 percent) was also reported in cardoon seeds subjected to foliar applications of *S. platensis* extracts. Again, the carotenoid, tocopherol, phenolic, and protein contents in grains harvested from wheat crops irrigated with 10 and 20 percent seawater were significantly increased in response to the application of water extracts of *S. maxima*. Figure 13 shows a representation of leaf and root colonization of tomato plant by cyanobacteria.

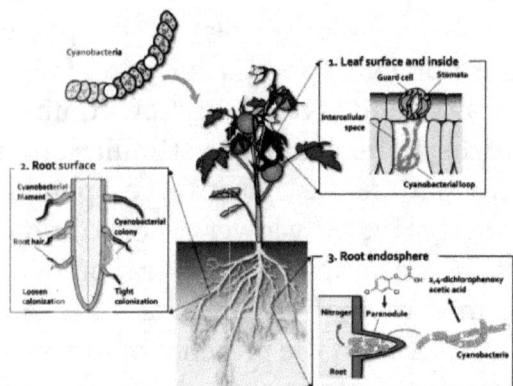

Fig. 13: Leaf and root colonization by cyanobacteria (Source: Lee and Ryu, 2021)

> *S. platensis* extracts have also been found useful in post-harvest treatments, and the formulation of edible coatings for fruits. The use of guar gum coating enriched with ethanolic extracts of S.

platensis to improve the chemical qualities and firmness during the storage of mango fruits has equally been reported. The study also showed that the phenolic content and radical scavenging activities in the treated fruits were increased. Other studies have reported the positive effects of microalgae-derived bio-stimulants on the leaves and concentrations of the bioactive substances in medicinal herbs such as peppermint.

Again, at low concentrations, algae extract has been shown to effectively stimulate plant germination and development. Algae extract are specifically able to enhance the sprouting of seeds, the growth of saplings, the production of photosynthetic pigment, and the ability of the seedlings to withstand external stressors. An Indian study investigated the effects of different concentrations of *S. platensis* extract on seed germination and seedling vigor of various crops. The parameters evaluated included seed germination percentage, and seedling vigor index-I and II. The results indicated significant variations in seed germination and seedling vigor among different crops, and concentrations of *S. platensis* extract. The 100 percent concentration of the extract however recorded the highest (85.75 percent) seed germination, indicating that such algal extracts could be used to enhance crop growth and productivity. A similar study in which spirulina powder and chickpea seeds were mixed at different concentrations was also reported to enhance the germination and growth of the seedlings.

Phycocyanin-rich spirulina extract (PRSE) has been used as bio-stimulation in hydroponically grown, vertical-farmed vegetables such as lettuce and cabbage. In such systems, PRSE application has been shown to reduce the time from seed to harvest by 6 days, increase the yield by 12 percent, and also improve the antioxidant and flavonoid levels in the vegetables. A 62 percent reduction in bacterial population and an overall increase in bacterial diversity have also been found in PRSE-treated hydroponic vegetable units. Presently, the global bio-stimulant market is one of the fastest-growing agriculture-related sectors, exhibiting an estimated compound annual growth rate (CAGR) of 10.65 percent in the period 2019 – 2027, in comparison to the much wider global inorganic fertilizer market, which is currently growing at a rate of 1.3 – 1.8 percent annually.

Spirulina as a Bio-Pesticide

Botanical pesticides are eco-friendly alternative plant-based products that utilize the different chemical compounds in plant tissues to inhibit the pests that attack crops. These compounds include terpenoids, steroids, phenols, coumarins, flavonoids, tannins, alkaloids, and cyanogenic glycosides, some of which are abundant in microalgae. The effect of *S. platensis* on insects has indeed been studied in numerous ways. For example, Aly and Abdou reported that 5 percent of *S. platensis* water extract increased larval mortality and malformation of the African cotton leafworm. The algae was utilized as a protein and carbohydrate source in phytophagous ladybug beetle, and showed effectiveness when applied as bio-pesticides against the 2nd and 4th instars of the black cutworm. Other studies have demonstrated the safety of *S. platensis as a* bio-control agent against the broad bean beetle, and its ability to stimulate germination and seedling growth of the plant.

In a recent study, Rashwan and Hammad investigated the toxic effects of water and ethanol extracts of *S. platensis and S. vulgar* as natural pesticides on the survival and biological parameters of the African cotton leafworm *S. littoralis,* as an eco-friendly and cost-effective initiative to protect different crops from economic damage, and losses caused by the insect, as well as reducing the risks resulting from excessive use of chemical pesticides on health and environment. The study demonstrated that treatment of 2nd and 4th larval instars of *S. littoralis* through feeding on leaf discs treated with water and ethanol extracts of tested algae was effective against the survival and biological parameters of the insect. *S. platensis* was specifically recommended as a promising natural alternative for controlling the cotton leaf worm and limiting the intensive use of chemical pesticides. The algae has also the added advantage of being a proven source of nutrients, and important phytochemicals for improving plant growth and yield. Thus, spirulina could be used as a bio-pesticide to decrease the pest populations in the farming environment, and as a bio-fertilizer and bio-stimulant for increasing crop yield and quality. An interesting study at King Saud University, Riyadh, Saudi Arabia, assessed the positive/negative impacts of *S. platensis* on almond moth, a serious pest of date fruits and grains under laboratory conditions. The algae powder was mixed with larvae diet at 0, 1, 2, 5, and 10 percent inclusion levels and fed to newly hatched larvae. The diet was replaced weekly, while data on the larvae were collected on alternate days. The spirulina powder was found to be a good source of nutrients at low inclusion levels, while at 5 and 10 percent levels it caused moderate and high mortality respectively (Figure 14).

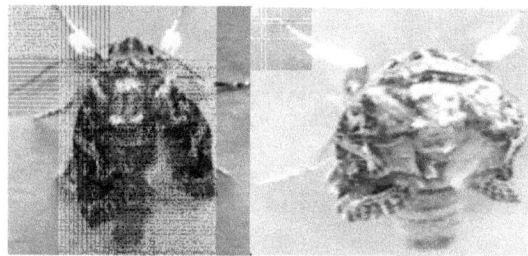

Fig. 14: Malformed and twisted wings of African cotton leaf worm treated with

T. platensis extract (Source: Rashwan and Hammad, 2020).

The lower formulation resulted in a shorter larval period compared to the 5 and 10 percent formulations that recorded 33 and 90 percent mortality respectively, with those reaching reaching adult stage and laying eggs. The study concluded that *S. platensis* could be used both as a nutritional supplement, and as a toxic substance to prevent the infestation of stored dates by the almond moth.

Conclusion

Spirulina can be used as a bio-fertilizer, bio-stimulant, and bio-pesticide in crop production. Crops treated with spirulina exhibit improved nutrient density, including higher concentrations of essential amino acids, vitamins, and antioxidants. As consumers increasingly prioritize sustainably sourced products, adoption of spirulina in agricultural practices could enhance marketability, and consumer trust in agricultural products. The knowledge about the potential value of spirulina in crop production is however limited or even lacking in many developing countries, and sub-Saharan Africa, even though it could be easily produced in most of these countries.

ENVIRONMENTAL AND OTHER USES OF SPIRULINA

Cultivation of spirulina for high yield (Image source: FFredom.com)

Prof Ifeanyi Charles Okoli

Introduction

Spirulina has gained significant attention for its versatile applications because of its dense nutrient profiles of proteins, polysaccharides, vitamins, minerals, and antioxidants, making it a potential candidate for a more sustainable biomass economy. The increasing global demand for plant-based renewable resources further amplifies the potential relevance of spirulina in many industries. Spirulina is being explored for its potential in bioremediation and wastewater treatment. Its ability to absorb heavy metals and toxins from contaminated water sources not only aids in clean-up efforts, but also provides a sustainable method for managing wastes. Additionally, its use in bioplastic production, and sustainable packaging presents an innovative approach to addressing the growing plastic pollution crisis. By leveraging its natural properties, industries can develop eco-friendly solutions that contribute to environmental sustainability. The natural pigments in spirulina are also being explored as food colorants and natural dyes, offering a sustainable alternative to non-sustainable synthetic options. Again, photosynthetic microalgae such as spirulina are proving excellent feedstock for biofuel production in the fast-developing biomass production technologies. The relatively high carbohydrate content of *S. platensis* which can reach 60 percent in the dry cell mass, and its abundant lipid content has particularly made spirulina a promising source of photosynthetic microbial biomass for biofuel production. The ongoing innovative research on several potential applications of spirulina points to its pivotal role across multiple sectors, thus making it an emerging cornerstone of the sustainable revolution.

The Use of Spirulina in Bioremediation of Polluted Sites

The current rapid global industrial development and urbanization have led to significant environmental changes resulting from the release of detrimental waste materials from these industries into the environment. The unprecedented discharge of industrial wastes, domestic wastewater, and atmospheric deposition into natural water has caused a series of toxic effects on both aquatic ecosystems and human health. The inappropriate release of pollutants such as heavy metals into the environment has therefore become a major global concern. Polluted terrestrial and aquatic ecosystems have been shown to affect the growth of living organisms when their concentrations exceed certain limits, depending on the type of pollutants. At low concentrations, some organisms may exhibit tolerance to the heavy metals, while at the same time accumulating them in their tissues. Recent bioremediation techniques offer opportunities for efficient clean-up of polluted water and land areas. Bioremediation is the process by which organic wastes are biologically degraded under controlled conditions to an innocuous state, or to levels below concentration limits established by regulatory authorities It is the use of living organisms, to degrade environmental contaminants into less toxic forms. For example, indigenous or imported microorganisms may be used to degrade or adsorb the contaminating compounds through reactions as a part of their metabolic processes.

Biosorption is the process of passive cation binding by dead or living biomass, representing a potentially cost-effective method of removing toxic heavy metals from industrial pollutant discharges and the environment. This approach could be employed most effectively in a concentration below 100 mg/L, where other techniques are relatively costly or ineffective. Different types of microorganisms have been shown to exhibit efficient metal biosorption capacity and are increasingly being employed in industrial cleanup operations due to the limitations associated with the traditional physicochemical methods for decontaminating polluted sites. Microalgae species having high protein, carbohydrate, and fat content, have been shown to interact with heavy metals through biosorption/bioaccumulation, leading to changes in the biomass composition, yield, and growth rate. Numerous studies have specifically shown that heavy metals such as lead, mercury, cadmium, and nickel interfere with the photosynthesis and enzymatic metabolism of algae, resulting in growth inhibition and even death. This is because alga as primary producers, and show high sensitivity to such heavy metals which indicates that they could serve as a test model for freshwater quality assessment and pollutant toxicity estimation.

Spirulina has specifically been recommended as an eco-friendly microalgae for bioremediation, nitrification, and carbon fixation. The key factors needed for optimal growth of spirulina are sufficient light intensity, a culture medium pH of about 9.5, a temperature of 30 to 35 °C, and availability of all the required nutrients. Spirulina cultivation requires minimal resources, since the algae utilizes carbon dioxide and sunlight to thrive, which makes it an effective tool for carbon sequestration and climate change mitigation. The living biomass has also the capacity to adsorb metal ions. In general, an algae species may exhibit high tolerance to some heavy metals and therefore could serve as biosorbent for those metals in bioremediation treatments. Heavy metals are non-biodegradable and cause a variety of disorders due to their bioaccumulation in living organisms, and are abundant in industrial effluents discharged into water bodies. Several studies have reported the ability of spirulina to selectively bioaccumulate some metals in the presence of others. Thus, the algae's natural capacity to absorb pollutants and improve water quality has been exploited for decades, making it an invaluable tool in environmental management. For example, live spirulina biomass in lead solutions of low concentration (below 50 mg/L) was shown to remove up to 74 percent of the metal in the first 12 minutes and 95 percent after 24 hours.

Spirulina has been effectively employed singly, and in combination with other microalgae in the adsorption and removal of heavy metals such as chromium (III), Cr(VI), and Copper(II) from wastewater. Rangsayatorn and colleagues reported that the uptake of the heavy metal by *S. platensis* was influenced by the pH of the solution, with the optimum level being 7 or neutral. They also reported a maximum adsorption capacity of 98.04 mg Cd per g biomass, while the uptake of heavy metal was rapid, with up to 78 percent of the metal sorption being completed within the first 5 minutes. Cardoso and colleagues reported that spirulina showed the highest removal efficiency of Bromine (Br), nitrate, nitrite, phosphate, sulfate, and chemical oxygen demand (COD) from fish pond wastewater, indicating its value in improving the environmental conditions and impacts of aquaculture production. Al-Dhabi reported that *S. platensis* is a better trace metals accumulator than *Chlorella vulgaris*, indicating its advantage in bio-remediation processes, although a liability when toxic metals are present. A study at the Technological University, Belagavi, Bangalore, India investigated the ability of chemically modified ginger and *S. maxima* powder to bioremediate hexavalent chromium from polluted lake water. The physicochemical parameters required for biosorption were optimized by batch adsorption experiments to attain maximum adsorption of up to 70 percent in the lake water Figure 15). The maximum adsorption obtained by ginger was 38, 78, and 88 percent, while for *S. maxima*, it was 22, 48, and 39 percent using natural, chemically modified, and immobilized biosorbents respectively in lake water. Treatment in lake water not only reduced the metal concentration but also mitigated the oxidative stress in the liver homogenates of zebrafish, depicting the effectiveness of the bioremediation process and the biosorbents used in the study.

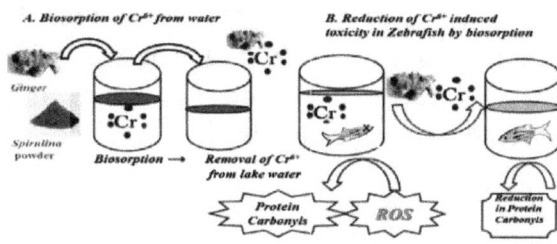

Fig. 15: The use of ginger and S. maxima as biosorbents in wastewater treatment (Source: Kurella et al., 2024)

These and several other studies inform the contemporary utilization of spirulina to combat pollution and restore ecosystems, positioning it as a key player in sustainable development initiatives to address climate change and environmental degradation. Integrating spirulina production into wastewater treatment processes can help in bioremediation efforts, extracting nutrients from waste while producing valuable biomass. This dual benefit aligns with the goals of various industries aiming to enhance sustainability while addressing environmental challenges.

Production of Biofuel from Spirulina

The increasing demand for fossil fuel and its corresponding depletion has in recent years driven the efforts to develop renewable energy sources with limited or no greenhouse gas emissions. Renewable energy sources are derived from biomass or biofuel. While energy-rich foods such as corn, soybeans, sugar cane, and cassava have been routinely used to produce biofuels such as bioethanol and biodiesel, this has raised several problems since it can increase global food prices and cause damage to the soil nutrient cycle. Microalgae on the other hand are rich in fat, carbohydrates, proteins, and other compounds that can be converted into biofuel through extractive and transesterification processes. Microalgae have several advantages over other biofuel sources since they are not in competition with human food, have higher photosynthetic efficiency, rapid growth, higher biomass yield, high energy value, and do not require large space for cultivation. Microalgae has also the added advantage of being used to produce several renewable fuels such as biodiesel, bioethanol, biohydrogen, methane, and syn-gas as shown in the framework in Figure 15.

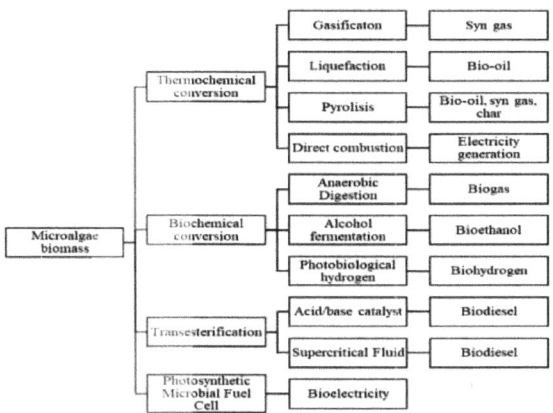

Fig. 15: Framework of microalgae biomass utilization as a renewable energy source (Source: Budhijanto et al., 2017)

Depending on the growth conditions, the carbohydrate content of *S. platensis* could reach up to 60 percent of the dry cell mass, and consists of polysaccharides, mostly in the form of starch, which are glucose monomers linked by α-1,4 glycosidic linkages. *S. platensis* intra-cellular carbohydrates can therefore be hydrolyzed and used for bioethanol production through fermentation with an ethanologenic microorganism. To obtain fermentable sugars from this biomass, a chemical, physical, or mechanical pretreatment is first carried out to break down the cells. Thereafter, the polysaccharides are hydrolyzed into fermentable sugars by enzymatic saccharification, and starch-type polysaccharides are cleaved into glucose units with the aid of enzymes such as α-amylase and glucoamylase. Endo-acting α-amylases are also employed to hydrolyze starch or glycogen into oligosaccharides, glucose, maltose, and maltodextrins. The cellulose content is, on the other hand, subjected to the action of cellulolytic enzymes to break down the cellulose into glucose molecules. Several reports have shown that it is possible to produce ethanol from the carbohydrates in *S. platensis* biomass, with the yields ranging from 34 to 93 percent from released sugars after carbohydrate hydrolysis of biomass, and subsequent fermentation with different bacteria and yeast, to achieve a volumetric productivity range of 0.14 to 1.0 gEtOH L^{-1} h^{-1}.

Rempel and colleagues produced bioethanol from *S. platensis* biomass using the saccharification process and thereafter, fermented the wastes from the bioethanol production to produce biomethane (Figure 16). Both, the enzymatic hydrolysis of the microalgae polysaccharides and the fermentation process recorded more than 80 percent efficiencies. The fermentation of the hydrolyzate into ethanol was possible without adding synthetic nutrients. Specifically, the direct conversion of the spirulina biomass into biomethane had an energy potential of 16,770 kJ.kg^{-1}, while bioethanol production from the hydrolyzed biomass presented 4,664 kJ.kg^{-1}. The sum of the energy potential obtained by producing bioethanol followed by biomethane production with the saccharification and fermentation of the residues was however 13,945 kJ.kg^{-1}. The results show that although the algae could be used to produce both biofuels, the direct production of biomethane is a more efficient approach, and that spirulina algae is a promising alternative renewable energy source. The actualization of these biofuel processes using microalgae is therefore a promising alternative approach to renewable energy production.

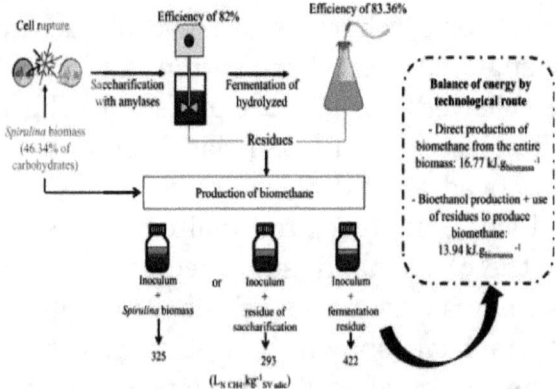

Fig. 16: The energy balance of the production of biomethane and bioethanol from spirulina biomass (Source: Rempel et al., 2019)

In a similar study, Werlang and her team investigated the effectiveness of pretreated and saccharified *S. platensis* biomass as a culture medium for ethanol production. The dried biomass was characterized, and the conditions for maximizing glucose release in the hydrolysate produced from enzymatic hydrolysis with amylases and cellulases were determined. The resulting slurry was thereafter used as a culture medium for ethanol production, using a metabolically engineered ethanologenic *E. coli* MS04. The result showed that 12.7 g L^{-1} ethanol was achieved after 9 hours of fermentation, which corresponded to a 92 percent conversion yield of the glucose content in the hydrolysate, 0.13 g of ethanol per 1 g of *Spirulina* biomass, and a volumetric productivity of 1.4 g of ethanol L^{-1} h^{-1}. The study therefore concluded that it is possible to obtain a high ethanol yield corresponding to 160 L per ton of dry biomass in a short time. Economic analysis of the use of microalgae as feedstock for biofuel production has however shown that the system could be designed to produce several bioproducts, thus improving the economic gains, and making the process commercially viable on large-scale. De Souza and colleagues demonstrated the relevance of bioethanol production in reinforcing microalgal production for agricultural, commercial, and industrial development. The initial pretreatment of the algae biomass with ethanologenic bacteria could therefore be exploited in the deployment of third-generation biorefineries similar to the ones used for lignocellulosic feedstock. By harnessing the energy stored in spirulina, industries can work towards reducing carbon footprints and promoting cleaner energy sources. The alga's capacity for carbon sequestration also positions it as a crucial player in climate change mitigation strategies.

Production of Bioplastic from Spirulina

The global volume of petrochemical-based plastics has been increasing over the decades, such that the total production is predicted to reach 33 billion tons by 2050, compared to the 0.28

billion tons produced in 2012. Inappropriate disposal strategies have resulted in large volumes of plastic waste accumulation in landfills, waterways, and oceans, causing significant hazards to human health and the environment. The chemical stability of common plastics which makes them amenable to numerous industrial applications also confers on them a long degradation time frame, thus allowing them to remain in the environment for many years, and constituting serious environmental and public health problems. Over the years, the search for suitable biodegradable alternatives to commodity plastics has resulted in the development of petrochemically-derived, biodegradable polymers such as polybutylene adipate terephthalate (PBAT), and biologically derived or biobased, non-degradable polymers such as polyethylene. These products have however been limited by either their petrochemical origin or their inability to readily degrade under natural conditions. Biobased and biodegradable polymers, such as the widely available polylactic acid (PLA) and poly(hydroxyalkanoate)s (PHAs), are being promoted as viable solutions to petrochemical-based plastic pollution. Bioplastics or organic plastics are derived from renewable biomass sources such as vegetable oil, corn starch, pea and cassava starch, and microalgae. Several microbial organisms including microalga produce the molecule polyhydroxybutyrate (PHB) which is a bioplastic as part of their food storage material. PHB has been detected in many cyanobacteria including spirulina where in the presence of a reduced carbon source, it constitutes up to 10 percent of the dry weight. The intra-cellular PHB accumulating property of *S. platensis* is higher than that of other spirulina species and some cyanobacteria. The major difference between PHB and petroleum-based plastics is the biodegradability of PHB when exposed to an environment populated by microorganisms such as bacteria, fungi, and algae. Under such conditions, the PHB is broken down to its essence-carbon dioxide and water recycled by the natural metabolic processes of these microbes.

Bioplastics are therefore biodegradable plastics that generally decompose into carbon dioxide, methane, water, and inorganic compounds through microbial processes. Three general approaches have been adopted in the production of bioplastics and include, polymers extracted directly from biomass either with or without modification; polymers produced with renewable raw materials and obtained using bio-intermediaries; and polymers produced directly by microorganisms in their natural or genetically modified state. Microalgae has been shown to serve as an excellent feedstock for bioplastic production due to its chemical composition, and many other advantages such as high yield and the ability to grow in a wide range of environments. The algal bioplastics are also easily degradable and can produce various forms of plastics including hybrid plastics, cellulose-based plastics, Poly-Lactic Acid (PLA), and bio-polyethylene. Generally, hybrid plastics are made by adding denatured algae biomass to petroleum-based plastics and polyethylene as fillers, while the PHB content of the algal biomass increases the biodegradability of the plastic. Biobased and biodegradable polymers are now readily available, with polylactic acid (PLA) and poly(hydroxyalkanoate)s (PHAs), being the most common ones that have the potential to reduce the dependency on petrochemical-based plastics and plastic pollution.

Zeller and colleagues used unmodified *S. platensis* and *C. vulgaris* cells to form bioplastics by subjecting them to heated compression molding, and demonstrated their thermoformability, although the bioplastics produced through this method had poor tensile strengths. Maheswari and Ahilandeswari also reported that the plasticizing, moldability, and biodegradable properties of S. platensis bioplastic are good due to its PHB content, although the cost is much higher than those of other chemical and organic polymers. Research on the precise mechanisms for transforming algal biomass into bioplastic is however ongoing and will need to address the poor mechanical properties reported in earlier studies which hinders their adoption as replacements for the high-volume commodity plastics. Iyer and colleagues in a recent report presented a fast and scalable method of producing strong and stiff backyard-compostable bioplastics from spirulina cells, without any binders or additives. Briefly, the approach involved subjecting the spirulina cells to conventional heat compression molding as shown in Figure 17.

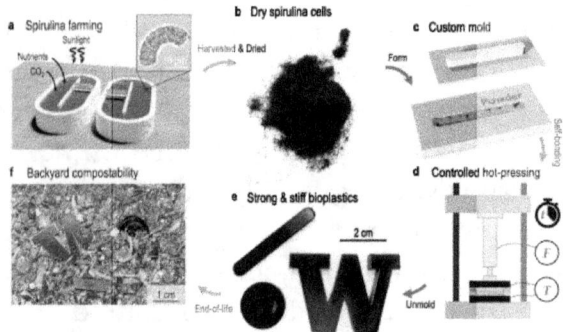

Fig. 17: Production of spirulina-based bioplastics (Source: Iyer et al., 2023)

The whole biomass powder was utilized in manufacturing without any extraction or chemical modification. Heat and pressure were applied at optimal conditions to compress the spirulina powder into rigid, thermo-formable bioplastics, which were amenable to further processing like a thermoplastic. The mechanical properties of the spirulina-based bioplastic were subjected to flexural tests by systematically varying the pressing time, temperature, and pressure, thereby making it possible to manipulate the mechanical properties of the bioplastics. The optimized bioplastics recorded modular strength comparable to or higher than commodity plastics, and surpassed those of thermoplastic starch and previously reported algal bioplastics. The spirulina-based bioplastic was compatible with existing polymer manufacturing infrastructure for backyard composting. Recent versions of bioplastic films from *S. platensis* have stronger tensile strength than commercial plastic bags, although their elongation at break is low. They are currently used as food, pharmaceutical, and cosmetic packaging materials. Bioplastic has also been produced from spirulina residue by adding polyvinyl alcohol. Those made from 95 percent Poly Lactic Acid (PLA) and 5 percent algae have been shown to have good mechanical properties, especially tensile strength and elongation at break, and can be decomposed in 45 days.

Production of Dyes from Spirulina

Before the advent of synthetic dyes, natural dyes derived from plants were used by humans in textile, leather, and other industries. The use of natural dyes however became unpopular due to several weaknesses such as the unavailability of dyes in standard ready-to-use forms, limited and non-reproducible colors, inability to meet huge industrial demands, and diminishing sources due to over-exploitation. Although readily available, synthetic dyes contain naphthol, vat dyestuffs, nitrates, acetic acid, soaping chemicals, enzymatic substrates, chromium-based materials, and heavy metals which when discharged with the effluent have been shown to constitute environmental pollutants. To overcome these problems, the focus has recently returned to eco-friendly materials, and safe production processes to maintain environmental and natural sustainability. As the interest in natural products continues to rise, spirulina's versatility has extended to cosmetics, food colorants, and textile dyes. The natural pigments found in spirulina can serve as eco-friendly alternatives to synthetic dyes for several industrial applications. The bio-pigments found in spirulina include chlorophyll, carotene, xanthophyll, and phycocyanin which contain blue pigments developed into varieties of food and cosmetic products. Only blue phycocyanin from spirulina is however used as a natural food colorant since there are other better sources of chlorophyll and carotenoids. Unmodified spirulina cells have also been used in additive manufacturing to create inks for direct ink writing. The printed structures were found to have mechanical properties and micromorphologies dependent on the drying method.

Spirulina dyes (Image source: Oterra.com)

Ciptandi and his team studied the potential of spirulina platensis as homemade natural dyes for the development of designs on textiles, and reported that the homemade dyes from spirulina paste matched the characteristics of the conventional handicraft industry. The dye was however found to have a low level of fastness resistance, and therefore, it is not suitable for dying products that require several washes. Agustina and colleagues determined the effect of adding variable concentrations (0.5, 1, and 1.5 percent) of spirulina residue as a natural dye on the characteristics of biocomposite films. The treated biocomposite films were analyzed for tensile strength and elongation at break, color, and morphology. The results showed that the 0.5 percent biocomposite recorded the highest tensile strength and elongation at break, while 1.5 percent biocomposite recorded the deepest greenish and yellowish colors, indicating that spirulina residue can be utilized as the plastic dyes. Spirulina extract stored for 10 days had an unstable color, although stability was reported in the absence of light. Micro-encapsulation has therefore been used to improve the antioxidant activity of natural dyes extracted from *S. platensis*.

Conclusion

The dense nutrient profiles of proteins, polysaccharides, vitamins, minerals, and antioxidants in spirulina makes the microalgae a potential candidate for exploitation in a sustainable biomass economy. Spirulina is therefore currently being explored for its potential in bioremediation, wastewater treatment, bioplastic, biofuel and natural dye production. Several ongoing innovative research on the applications of spirulina points to its potential roles across multiple sectors, thus making it an emerging natural eco-friendly resource.

Bibliographic References

Agustini,T.W., Ma'ruf, W.F., Widayat, Wibowo, B.A. and Hadiyanto (2017). Study on the effect of different concentration of *Spirulina platensis* paste added into dried noodle to its quality characteristics. *Proc. IOP Conf. Series: Earth and Environmental Science,* 55. 012068 doi:10.1088/1755-1315/55/1/012068

Al-Dhabi, N.A. (2013). Heavy metal analysis in commercial spirulina products for human consumption. *Saudi J. Biol. Sci.,* 20, 383–388.

Aleksandrovna, G.G., Viktorovna, N.L. and Dementievna, Z.I. (2019). Spirulina as a protein ingredient in a sports nutrition drink. *Advances in Health Sciences Research,* 17, 162 - 165.

Al Fadhly, N.K.Z., Alhelfi, N., Altemimi, A.B., Verma, D.K., Cacciola, F., Narayanankutty, A. (2022). Trends and technological advancements in the possible food applications of spirulina and their health benefits: A Review. *Molecules,* 27, 5584. https://doi.org/10.3390/molecules27175584

Ali, S.K. and Saleh, A.M. (2012). Spirulina - an overview. *International Journal of Pharmacy and Pharmaceutical Sciences,* 4(3), 9 -15.

Aly, M.S. and Abdou, W.L. (2010). The effect of native *Spirulina platensis* on the developmental biology of *Spodoptera littoralis* (Boisd). *J. Genet. Eng. Biotechnol.,* 8, 65 - 70.

Anuradha, V. and Vidhya, D. (2001). Impact of administration of Spirulina on the blood glucose levels of selected diabetic patients. *Indian J. Nutri. Dietetics,* 38, 40–44.

Atik, D., Gurbuz, B., Boluk, E. and Palabiyik, I. (2021). Development of vegan kefir fortified with *Sprulina platensis*. *Food Bioscience,* 42, 101050.

Ayala, F. (1998). Guía sobre el cultivo de Spirulina. In: *Biotecnología de microorganismos fotoautótrofos.* Motril, Granada, España. Pp, 3–

20.

Ayehunie S, Belay A, Baba TW, Ruprecht RM (1998). Inhibition of HIV-1 replication by an aqueous extract of *Spirulina platensis* (*Arthrospira platensis*)
J Acq Immune Def Syndromes Human Retrovirol 18(1), 7 - 12.

Barrón, L.B., Torres-Valencia, M.J., Chamorro-Cevallos, G. and Zúñiga-Estrada, A. (2008). Spirulina as an antiviral agent. In: M.E. Gershwin & A. Belay (eds.). *Spirulina in human nutrition and health*, CRC Pres, Taylor and Francis. Pp, 227.

Bhattacharya, S. and Shivaprakash, M.K. (2005). Evaluation of three Spirulina species grown under similar conditions for their growth and biochemicals. *J. Sci. Food Agric.*, 8, 333–336.

Chaouachi, M., Gautier, S., Carnot, Y., Guillemot, P., Pincemail, J., Moison, Y., *et al.* Spirulina supplementation prevents exercise-induced lipid peroxidation, inflammation and skeletal muscle damage in elite rugby players. *J Hum Nutr Diet,* 35, 1151–63. doi: 10.1111/jhn.13014

Colla, L.M., Muccillo-Baisch, A.L. and Costa, J.A.V. (2008). *Spirulina platensis* effects on the levels of total cholesterol, HDL and triacylglycerols in rabbits fed with a hypercholesterolemic diet. *Braz. Arch. Biol. Technol., 51*, 405–411.

Cardoso, L., Duarte, J.H., Costa, J.A.V., Assis, D.J., Lemos, P.V.F., Druzian, J.I. et al. (2020). Spirulina sp. as a bioremediation agent for aquaculture wastewater: Production of high added value compounds and estimation of theoretical biodiesel. *BioEnergy Research,* 14(1), 254-264.

Costa, J.A.V., Freitas, B.C.B., Rosa, G.M., Moraes, L., Morais, M.G. and Mitchell, B.G. (2019). Operational and economic aspects of Spirulina-based biorefinery. *Bioresource Technology,* 292 (2019), 121946.

Dataintelo (2022). Global spirulina market by type, application and by region forecast from 2022 to 2030. dataintelo.com

Dalle Zotte, A., Sartori, A., Bohatir, P., Remignon, H. and Ricci, R. (2013). Effect of dietary supplementation of spirulina (*Arthrospira platensis*) and thyme (*Thymus vulgaris*) on growth performance, apparent digestibility and health status of companion dwarf rabbits. *Livest. Sci.,* 152, 182-191.

De Souza, M.P., Hoeltz, M., Gressler, P.D., Benitez, L.B. and Schneider, R.C.S. (2018). Potential of microalgal bioproducts: General perspectives and main challenges. *Waste Biomass Valorization,* 10(8), 2139–2156.

El-Sakhawy, M.A., Iqbal, M.Z., Gabr, G.A., Alqasem, A.A., Ateya, A.A.E., Ahmed, F.A., El-Hashash, S.A., Ibrahim, H.S., El-Ghiet, U.M.A (2023). The mechanism of action of spirulina as antidiabetic: A narrative review. *Italian Journal of Medicine,* 17, 1639.

El-Sheekh, M.M., Daboo, S., Swelim, M.A. and Mohamed, S. (2014). Production and characterization of antimicrobial active substance from Spirulina platensis," *Iranian Journal of Microbiology,* 6(2), 112–119.

Evans, A.M., Smith, D.L. and Moritz, J.S. (2015). Effects of algae incorporation into broiler starter diet formulations on nutrient digestibility and 3 to 21 day bird performance. *J. Appl. Poultry Res.,* 24(2), 206-214.

FAO (2008). A review on culture, production and use of spirulina as food for humans and feeds for domestic animals and fish. FAO Fisheries and Aquaculture Circular No. 1034. Food and Agriculture Organization of the United Nations, Rome.

Finamore, A., Palmery, M., Bensehaila, S. and Peluso, I. (2017). Antioxidant, immunomodulating, and microbial-modulating activities of the sustainable and ecofriendly *spirulina. Oxidative Medicine and Cellular Longevity,* 2017, Article ID 3247528, 14 pages https://doi.org/10.1155/2017/3247528

Gharib, F.E. and Ahmed, E.Z. (2023). *Spirulina platensis* improves growth, oil content, and antioxidant activity of rosemary plant under cadmium and lead stress. *Scientific Report*, (2023) 13: 8008 https://doi.org/10.1038/s41598-023-35063-1

Ghazal, G.A. (2016). Nutritional characteristics and bioactive compounds of different ovo-vegetarian diets supplemented with spirulina. *Annals of Agric. Sci., Moshtohor*, 54(2), 307 – 322.

Gogna, S., Kaur, J., Sharma, K., Prasad, R., Singh, J., Bhadariya, V., Kumar, P. and Jarial, S. (2022). Spirulina- An edible cyanobacterium with potential therapeutic health benefits and toxicological consequences. *Journal of the American Nutrition Association*, DOI: 10.1080/27697061.2022.2103852

Ferrazzano, G.F., Papa, C., Pollio, A., Ingenito, A., Sangianantoni, G. and Cantile, T. (2020). Cyanobacteria and microalgae as sources of functional foods to improve human general and oral health. *Molecules* 2020, *25*, 5164; doi:10.3390/molecules25215164

Grahl, S., Strack, M., Weinrich, R. and Morlein, D. (2018). Consumer-oriented product development: The Conceptualization of novel food products based on spirulina (*Arthrospira platensis*) and resulting consumer expectations. *Journal of Food Quality*, 1919482, 11 pages https://doi.org/10.1155/2018/1919482

Gautam, L., Chaturvedi, N. and Gupta, A. (2015). Development of micro-nutrients rich homemade extruded food products with the incorporation of processed foxtail millet, wheat and chickpea. *Indian Journal of Community Health*, 26, 288 - 293.

GlobeNewswire (2022). Spirulina market analysis report and region forcast, 2022 - 2030. www.globenewswire.com

Gonçalves, A.L. (2022). The Use of microalgae and cyanobacteria in the improvement of agricultural practices: A review on their biofertilising, biostimulating and biopesticide roles. *Appl. Sci., 11*, 871.

Grinstead, G.S., Tokach, M.D., Dritz, S.S., Goodband, R.D. and Nelssen, J.L. (2000). Effects of *Spirulina platensis* on growth performance of weanling pigs. *Anim. Feed Sci. Technol.*, 83, 237–247.

Guan, F., Fu, G., Ma, Y., Zhou, L., Li, G., Sun, C. and Zhang, T. (2024). *Spirulina platensis* alleviates chronic inflammation with modulation of gut microbiota and intestinal permeability in rats fed a high-fat diet. *Journal of Functional Foods*, 116 (2024) 106158

Guldas, M., Gurbuz, O., Cakmak, I., Yildiz, E.L.I.F., Sen, H. (2002). Effects of honey enrichment with *Spirulina platensis* on phenolics, bioaccessibility, antioxidant capacity and fatty acids. *LWT-Food Sci. Technol.*, 153, 112461.

Gunes, S., Tamburaci, S., Dalay, M.C. and Deliloglu, G.I. (217). In vitro evaluation of Spirulina platensis extract incorporated skin cream with its wound healing and antioxidant activities. *Pharm. Biol.*, 55, 1824–1832.

Gouveia, L., Coutinho, C., Mendonça, E., Batista, A.P., Sousa, I., Bandarra, N.M. and Raymundo, A. (2008). Functional biscuits with PUFA-ω3 from *Isochrysis galbana*. *J. Sci. Food Agric*, 88, 891 – 896.

Grawish, M.E., Zaher, A.R., Gaafar, A.I. and Nasif, W.A. (2010). Long-term effect of Spirulina platensis extract on DMBA-induced hamster buccal pouch carcinogenesis (immunohistochemical study). *Med. Oncol.*, 27, 20–28.

Habib, M.A.B., Parvin, M., Huntington, T.C. and Hasan, R.M. (2008). A Review on culture, production and use of spirulina as food for humans and feeds for domestic animals and fish. *Fisheries and Aquaculture Circular,* No. 1034. Food and Agricultural Organization, Rome, Italy.

Hardiningtyas, S.D., Febby Amanda Putri, F.A. and Setyaningsih, I. (2022) Antibacterial activity of ethanolic *Spirulina platensis* extractwater soluble chitosan nanoparticles. *IOP Conference Series: Earth and Environmental Science,* 1033 (2022) 012053 doi:10.1088/1755-1315/1033/1/012053

Henrikson, R. (1989). *Earth food spirulina.* San Rafael, California, USA, Ronorc Enterprises, Inc.

Hayashi, T., Hayashi, K., Maeda, M. and Kojima, I. (1996). Calcium Spirulan, an inhibitor of enveloped virus replication, from a blue-green alga *Spirulina platensis. J. Nat. Prod.,* 59(1) 83-87

Huang, H., Liao, D., Pu, R.. and Cui, Y. (2018). Quantifying the effects of spirulina supplementation on plasma lipid and glucose concentrations, body weight, and blood pressure. *Diabetes Metab. Syndr. Obes., 11,* 729.

Hwang, Y., Kwang-Jin, K., Su-Jin, K., Seul-Ki, M., Seong-Gyeol, H., Young-Jin, S. and Sung-Tae, Y. (2018). Suppression effect of astaxanthin on osteoclast formation in vitro and bone loss *in vivo. Int. J. Mol. Sci., 19,* 912.

Iyer, H., Grandgeorge, P., Jimenez, A.M., Campbell, I.R., Parker, M., Holden, M., Venkatesh, M., Nelsen, M., Nguyen, B. and Roumeli, E. (2023). Fabricating strong and stiff bioplastics from whole spirulina cells. *Advanced Functional Materials,* 2023, 2302067. DOI: 10.1002/adfm.202302067

Jagdale, P.E. (2012). Production of value added extruded products of spirulina to outwitted the malnutrition- A comprehensive study. *International Journal of Botany Studies,* 6(5), 598 - 602.

Jung, F., Krüger-Genge, A., Waldeck, P. and Küpper, J.-H. (2019). *Spirulina platensis,* a super food? *J. Cell. Biotechnol., 5,* 43–54.

Kalafati M., *et al.* (2010). Ergogenic and antioxidant effects of spirulina supplementation in humans. *Medicine and Science in Sports and Exercise,* 42(1), 142-151. https://www.ncbi.nlm.nih.gov/pubmed/20010119

Khalid, S., Chaudhary, K., Aziz, H., Amin, S., Sipra, H.M., Ansar,

S., Rasheed, H., Naeem, M. and Onyeka, (2024). Trends in extracting protein from micro-algae *Spirulina platensis*, using innovative extraction techniques: Mechanisms, potentials, and limitations. *Critical Review in Food Science and Nutrition*, 1 - 17. doi.org/10.1080/10408398.2024

Khan, M., Shobha, J.C., Mohan, I.K., Naidu, M.U., Sundaram, C., Singh, S., Kuppusamy, P. and Kutala, V.K. (2005). Protective effect of Spirulina against doxorubicin-induced cardio toxicity *Phytother. Res.*, 19(12), 1030-1037.

Khan, M., Varadharaj, S., Shobba, J.C., Naidu, M.U., Parinandi, N.L., Kutala, V.K. and Kuppusamy, P. (2006). C-Phycocyanin ameliorates doxorubicin-induced oxidative stress and apoptosis in adult rat cardiomyocytes. *J. Cardiovasc. Pharmacol.* 47(1), 9-20.

Koníčková, R., Vanková, K., Vaníková, J., Vánová, K., Muchová, L., Subhanová, I., Zadinová, M., Zelenka, J., Dvorák, A., Kolár, M., Strnad, H., Rimpelová, S., Ruml, T., Wong, R.J. and Vítek, L. (2014). Anti-cancer effects of blue-green alga *Spirulina platensis*, a natural source of bilirubin-like tetrapyrrolic compounds. *Annals of Hepetology*, 13(2), 273 - 283.

Koru, E. (2012). Earth food spirulina (Arthrospira): In: Yehia El-Samragy (Ed.). *Production and quality standards, food additive.* InTech, Available from: http://www.intechopen.com/books/food-additive/earth-food-spirulina-arthrospira-production-and-qualitystandart

Kurella1, B.R., Ramanujan, A.S., Nagaraj, B., Narayanappa, R. and Kunne, S.J. (2024). Chemically modified ginger and spirulina for bioremediation of hexavalent chromium from polluted water. *Journal of Integrated Science and Technology*, 12(4), 782

Lia Longodor, A., Coroian, A., Balta, I., Taulescu, M., Toma, C., Sevastre, B., Marchis, Z., Andronie, L., Pop, I., Matei, F., *et al.* (2021). Protective effects of dietary supplement spirulina (*Spirulina platensis*) against toxically impacts of monosodium glutamate in blood and behavior of *Swiss mouse. Separations, 8*, 218. https://doi.org/10.3390/separations8110218

Lakshmi, E., Tamilselvi, S., Priya, V. and Author, C. (2017). Spirulina: A contemporary food supplement. *International Journal of Food,* 5(3), 67-71.

Layam, A., Gannavarapu, S.B. and Pilla, K. (2016). Effect of supplementation of *Spirulina platensis* to enhance the zinc status in plants of *Amaranthus gangeticus, Phaseolus aureus* and *tomato. Advances in Bioscience and Biotechnology,* 7, 289-299.

Li, T-T, Tong, A-J., Liu Y-Y, et al. (2019). Polyunsaturated fatty acids from microalgae *Spirulina platensis* modulates lipid metabolism disorders and gut microbiota in high-fat diet rats. *Food Chem Toxicol.*, 2019, 131, 110558.

Lyons, C.L., Kennedy, E.B. and Roche, H.M. (2016). Metabolic inflammation-differential modulation by dietary constituents. *Nutrients,* 2016, 8, 247.

Machihara, K., Oki, S., Maejima, Y., Koseki, Y., Imai, Y. and Namba, T. (2023). Restoration of mitochondrial function by spirulina polysaccharide via upregulated SOD2 in aging fibroblasts. *iScience,* 26, 107113 July 21, 2023 ª 2023 The Authors. https://doi.org/10.1016/j.isci.2023.107113

Maheswari, N.U. and Ahilandeswari, K. (2011). Production of bioplastic using Spirulina platensis and comparison with commercial plastic. *Research in Environment and Life Sciences,* 4(3), 133-136.

Martins, C.F., Assunção, J.P., Santos, D.M.R., Madeira, M.S.M.S., Alfaia, C.M.R.P.M., Lopes, P.A.A.B., Coelho, D.F.M., Lemos, J.P.C., de Almeida, A.M., Prates, J.A.M. and Freire, J.P.B. (2021). Effect of dietary inclusion of Spirulina on production performance, nutrient digestibility and meat quality traits in post-weaning piglets. *J Anim Physiol Anim Nutr.,* 105, 247–259.

Masten Rutar, J., Jagodic Hudobivnik, M., Nečemer, M., Vogel Mikuš, K., Arčon, I. and Ogrinc, N. (2022). Nutritional quality and safety of the spirulina dietary supplements sold on the Slovenian market. *Foods,* 2022, 11, 849.

McCarty, M.F. and Kerna, N.A. (2021). Spirulina rising: The application of microalgae in protecting human health and treating disease. *EC Nutrition,* 16(7), 141-152.

Morsy, O.M., Sharoba, A.M., EL-Desouky, A.I., Bahlol, H.E.M. and Abd El Mawla, E.M. (2014). Production and evaluation of some extruded food products using spirulina algae. *Annals of Agric. Sci., Moshtohor,* 52(4), 329 – 342.

Nawar Dalia, A.S. and Ibraheim Sabreen Kh, A. (2014). Effect of algae extract and nitrogen fertilizer rates on growth and productivity of peas. *Middle East Journal of Agriculture Research,* 3(4), 1232–1241. ISSN 2077–4605

Neumann, C., Velten, S. and Liebert, F.N. (2018). Balance studies emphasize the superior protein quality of pig diets at high inclusion level of algae meal (*Spirulina platensis*) or insect meal (*Hermetia illucens*) when adequate amino acid supplementation is ensured. *Animals,* 2018, 8, 172.

Neyrinck, Audrey M., Taminiau, B., Walgrave, H., Daube, G., Cani, P.D., et. al. (2017). *Spirulina protects against hepatic inflammation in aging: An effect related to the modulation of the gut microbiota?.* Nutrients, 9(6), 633 [1-13 http://hdl.handle.net/2078.1/185760 -- DOI:10.3390/nu9060633

Park, J.H., Lee, S.I. and Kim, I.H. (2018). Effect of dietary *Spirulina* (Arthrospira) *platensis* on the growth performance, antioxidant enzyme activity, nutrient digestibility, cecal microflora, excreta noxious gas emission, and breast meat quality of broiler chickens. *Poult. Sci.,* 97, 2451–2459.

Parvin, M. (2006). Culture and growth performance of *Spirulina platensis* in supernatant of digested poultry waste. M.S. Thesis, Bangladesh Agricultural University, Mymensingh, Bangladesh.

Pabon, M.M., Jernberg, J.N., Morganti, J., Contreras, J., Hudson, C.E., *et al.* (2012). A spirulina-enhanced diet provides neuroprotection in an a-synuclein model of Parkinson's disease. *PLoS ONE,* 7(9), e45256. doi:10.1371/journal.pone.0045256

Piccolo, A. and Short, C. (Undated). Spirulina production in ESA-10region- The way forward.Smart Fiche, 4, 1 - 4.

Piovan, A., Battaglia, J., Filippini, R., Dalla Costa, V., Facci, L., Argentini, C., Pagetta, A., Giusti, P. and Zusso, M. (2021). Pre- and early posttreatment with *Arthrospira platensis* (Spirulina) extract impedes lipopolysaccharide triggered neuroinflammation in microglia. *Front. Pharmacol.,* 12, 724993.

Priyanka, S., Varsha, R., Verma, R. and Babu, A.S. (2023). Spirulina: A spotlight on its nutraceutical properties and food processing applications.*J. Microbiol. Biotech. Food Sci., 20xx : x (x) e4785*

Rangsayatorn, N., Upatham, E.S., Kruatrachue, M., Pokethitiyook, P. and Lanza, G.R. (2002). Phytoremediation potential of *Spirulina* (*Arthrospira*) *platensis*: Biosorption and toxicity studies of cadmium. *Environmental Pollution,* 119(1), 45–53.

Rashwana, R.S. and Hammad, D.M. (2020). Toxic effect of *Spirulina platensis* and *Sargassum vulgar* as natural pesticides on survival and biological characteristics of cotton leaf worm *Spodoptera littoralis. Scientific African,* 8 (2020), e00323

Rempela, A., Sossellaa, F.S., Margaritesa, A.C., Astolfib, A.L., Steinmetzc, R.L.R., Kunzc, A., Treicheld, H. and Colla, L.M. (2019). Bioethanol from *Spirulina platensis* biomass and the use of residuals to produce biomethane: An energy efficient approach. *Bioresource Technology,* 288 (2019) 121588

Rezaiyan, M., Sasani, N., Kazemi, A., Mohsenpour, M.A., Babajafari, S., Mazloomi, S.M., Clark, C.C.T., Hematyar, J., Far, Z.G., Azadian,

M. and Zareifard, A. (2023). The effect of spirulina sauce on glycemic index, lipid profile, and oxidative stress in type 2 diabetic patients: A randomized double-blind clinical trial. *Food Science and Nutrition*, 11, 5199–5208.

Roohani, A., Abedian, K., Kenari, A.A., Kapoorchali, M.F., Bourani, M.S., Zorriehzahra, M.J. et al. (2019). Effect of *Spirulina platensis* as a complementary ingredient to reduce dietary fish meal on the growth performance, whole-body composition, fatty acid and amino acid profiles, and pigmentation of Caspian brown trout (*Salmo trutta caspius*) juveniles. *Aquaculture Nutrition*, 2019, 1- 13.

Saharan, V. and Jood, S. (2017). Nutritional composition of *Spirulina platensis* powder and its acceptability in food products. *Int. J. Adv. Res.*, 2017, 5, 2295–2300.

Sánchez, M., Castillo, B.J., Rozo, C. and Rodríguez, I. (2003). Spirulina (Arthrospira): An edible microorganism. A review. *Universitas Scientiarum*, 8(1), 1-16.

Sanchez-Oliver, A.J., Contreras-Calderon, J., Puya-Braza, J.M. and Guerra-Hemandez, E. Quality analysis of commercial protein powder supplements and relation to characteristics declared by manufacturer. *Food science and Technology*, 97, 100 - 108.

Satyaraj, E., Reynolds, A., Engler, R., Labuda, J. and Sun, P. (2021). Supplementation of diets with spirulina influences immune and gut function in dogs. *Front. Nutr.*, 8, 667072. doi: 10.3389/fnut.2021.667072

Shalaby, E.A., Shanab, S.M.M. and Singh, V. (2010). Salt stress enhancement of antioxidant and antiviral efficiency of *Spirulina platensis*. *Journal of Medicinal Plants Research*, 4(24), 2622 - 2632.

Siedenburg, J.R. and Cauchi, J.P. (2022). Spirulina (Arthrospira Spp) as A complementary Covid-19 response option: Early evidence of promise. *Current Research in Nutrition and Food Science*, 10 (1), 129-144.

Siva Kiran, R.R., Madhu, G.M. and Satyanarayana, S.V. (2015). Spirulina in combating Protein Energy Malnutrition (PEM) and Protein Energy Wasting (PEW)—a review. *Journal of Nutrition Research,* 3(1), 62 – 79.

Soni, R.A., Sudhakar, K. and Rana, R.S. (2017). Spirulina-from growth to nutritional product: A review. *Trends in Food Science and Technology,* 69, 157 - 171.

Sow, S. and Ranjan, S. (2021). Cultivation of spirulina: An innovative approach to boost up agricultural productivity. *The Pharma Innovation Journal,* 10(3), 799 - 813.

Szmejda, K., Duli ́nski, R., Byczy ́nski, L., Karbowski, A., Florczak, T. and Zyła, K. (2018). Analysis of the selected antioxidant compounds in ice cream supplemented with *Spirulina* (*Arthrospira platensis*) extract. *Biotechnol. Food Sci., 82,* 41 – 48.

Tanticharoen, M., Reungjitchachawali, M., Boonag, B., Vonktaveesuk, P., Vonshak, A. and Cohen, Z. (1994). Optimization of gamma-linolenic acid (GLA) production in *Spirulina platensis. J. Appl. Phycol.,* 6, 295 – 300.

Teas, J., Hebert, J.R., Fitton, J.H. and Zimba, P.V. (2004). Algae - a poor man's HAART? *Med Hypothesis,* 64(4), 507-510.

Tidjani, A., Doutoum, A.A., Alio, H.M. and Aguid, M.N. (2018). Study of the microbiological quality of improved spirulina marketed in Chad. *Nutrition and Food Science International Journal,* 4(4), 1 - 5.

Villaro-Cos, S., Sanchez, J.L.G., Acien, G. and Lafarge, T. (2024). Research trends and current requirements and challenges in the industrial production of spirulina as food source. *Trends in Food Science and Technology,* 143, 104280

Yousefi, H. and Douna, B.K. (2023). Risk of nitrate residues in food products and drinking water. *Asian Pac. J. Environ. Cancer,* 6, 69–79.

Yousefi, R., Mottaghi, A. and Saidpour, A. (2018). *Spirulina platensis* effectively ameliorates anthropometric measurements and obesity-related metabolic disorders in obese or overweight healthy individuals: a randomized controlled trial. *Complement Ther. Med.*, 40, 106–12.

Wu, D., Nie, P., Cuello, J., He, Y., Wang, Z. and Wu, H. (2011). Application of visible and near infrared spectroscopy for rapid and non-invasive quantification of common adulterants in spirulina powder. *J. Food Eng.*, 102, 278 – 286.

Ting Yu, T., Wang, Y., Chen, X., Xiong, X., Tang, Y. and Lin, L. (2020). *Spirulina platensis* alleviates chronic inflammation with modulation of gut microbiota and intestinal permeability in rats fed a high-fat diet. *Journal of Cellular and Molecular Medicine*, 24, 8603–8613.

Werlang, E.B., Julich, J., Muller, M.V.G., Neves, F.F., Sierra-Ibarra, E., Martinez, A. and Schneider, R.C.S. (2020). Bioethanol from hydrolyzed Spirulina (*Arthrospira platensis*) biomass using ethanologenic bacteria. *Bioresources and Bioprocessing*, (2020) 7, 27.

Zarrouk, C. (1966). Contribution à l'étude d'une cyanophycée influencée de divers facteurs physiques et chimiques sur la croissance et la photosynthèse de *Spirulina maxima* (Setch. et Gardner) Geitler. PhD Thesis, University of Paris, Paris, France.

Zeinnalian, R., Farhangi, M.A. and Shariat, A. (2017). The effects of *Spirulina platensis* on anthropometric indices, appetite, lipid profile and serum vascular endothelial growth factor (VEGF) in obese individuals: A randomized double blinded placebo controlled trial. *BMC Complement Altern Med.*, 17 (1), 225 - 238.

Zeller, M.A., Hunt, R., Jones, A. and Sharma, S. (2013). *J. Appl. Polym. Sci.*, 130, 3263.

ABOUT THE AUTHOR

Okoli, Ifeanyi Charles is a Professor of animal production, and health at the Federal University of Technology Owerri, Nigeria. He is the coordinator of the consortium on Nigerian Poultry Feeds Research and Development (NIPOFERD). He is a Science Technology Innovation (STI) facilitator, academic journal publisher, tropical animal health, and production consultant. He writes regularly on the Research Tropica blog, an open-source information platform for communicating current tropical research trends to both science and non-science individuals.